3rd Iberian Meeting on Aerosol Science and Technology (RICTA 2015)

IOP Conference Series: Earth and Environmental Science
Volume 28

Elche, Spain
29 June – 1 July 2015

ISBN: 978-1-5108-1633-6
ISSN: 1755-1307

Printed from e-media with permission by:

Curran Associates, Inc.
57 Morehouse Lane
Red Hook, NY 12571

Some format issues inherent in the e-media version may also appear in this print version.

Printed by Curran Associates, Inc. (2016)

For permission requests, please contact the Institute of Physics at the address below.

Institute of Physics
Dirac House, Temple Back
Bristol BS1 6BE UK

Phone: 44 1 17 929 7481
Fax: 44 1 17 920 0979

techtracking@iop.org

Additional copies of this publication are available from:

Curran Associates, Inc.
57 Morehouse Lane
Red Hook, NY 12571 USA
Phone: 845-758-0400
Fax: 845-758-2634
Email: curran@proceedings.com
Web: www.proceedings.com

Table of contents

Volume 28

3rd Iberian Meeting on Aerosol Science and Technology (RICTA 2015)
29 June to 1 July 2015, Elche, Spain

Accepted papers received: 10 November 2015
Published online: 1 December 2015

Preface

Papers

IOP Conf. Series: Earth and Environmental Science **28** (2015) 011001 doi:10.1088/1755-1315/28/1/011001

Preface

The Third Iberian Meeting on Aerosol Science and Technology (RICTA 2015) was held in the city of Elche (province of Alicante, Spain) from 29 June to 1 July 2015, at Centro de Congresos Ciutat d'Elx. This event was organized and hosted by the Statistical and Computational Physics Laboratory (SCOLAb) of Universidad Miguel Hernández under the auspices of AECyTA, the Spanish Association for Aerosol Science and Technology Research. As in previous editions, the participation of young researchers was especially welcome, with the organization of the VI Summer School on Aerosol Science and Technology and awards for the best poster and PhD thesis, in recognition of outstanding research or presentations focusing on aerosols, during the early stage of their scientific career.

RICTA 2015 aims to present the latest research and advances on the field of aerosols, as well as fostering interaction among the Portuguese and Spanish communities. The meeting gathered over 70 participants from 7 different countries, covering a wide range of aerosol science and technology. It included invited lectures, keynote talks, and several specialized sessions on different issues related to atmospheric aerosols, radiation, instrumentation, fundamental aerosol science, bioaerosols and health effects.

The editors would like to express their sincere gratitude to all the participants, in particular, those who contributed to this special issue by submitting their papers to convey the current science discussed at RICTA 2015. In this special issue a series of peer-reviewed papers that cover a wide range of topics are presented: aerosol formation, emission, as well as aerosol composition in terms of physical and optical properties, spatial/temporal distribution of aerosol parameters, aerosol modeling and atmospheric effects, as well as instrumentation devoted to aerosol measurements. Finally, we also thank the referees for their valuable revision of these papers.

IOP Conf. Series: Earth and Environmental Science **28** (2015) 011002 doi:10.1088/1755-1315/28/1/011002

Peer review statement

All papers published in this volume of *IOP Conference Series: Earth and Environmental Science* have been peer reviewed through processes administered by the proceedings Editors. Reviews were conducted by expert referees to the professional and scientific standards expected of a proceedings journal published by IOP Publishing.

IOP Conf. Series: Earth and Environmental Science **28** (2015) 012001 doi:10.1088/1755-1315/28/1/012001

Atmospheric Particle Size Distributions in the Spanish Network of Environmental DMAs (REDMAAS)

E Alonso-Blanco[1], F J Gómez-Moreno[1], B Artíñano[1], S Iglesias Samitier[2], V Juncal[2], M Piñeiro Iglesias[2], P López Mahía[2], N Pérez[3], M Brines[3], A Alastuey[3], X Querol[3], B A de la Morena[4], M I García[5,6], S Rodríguez[5], M Sorribas[7,8], G Titos[7,8], H Lyamani[7,8] and L Alados-Arboledas[7,8]

[1] Department of Environment Associated Unit CSIC-CIEMAT on Atmospheric Pollution, CIEMAT, Madrid, Spain.
[2]Grupo Química Analítica Aplicada, Instituto Universitario de Medio Ambiente (IUMA), Centro de Investigaciones Científicas Avanzadas (CICA), Departamento de Química Analítica, Facultade de Ciencias, Universidade da Coruña, A Coruña, Spain.
[3]Institute of Environmental Assessment and Water Research (IDAEA-CSIC), Barcelona, Spain.
[4]Atmospheric Sounding Station "El Arenosillo", INTA, Mazagón-Huelva, Spain.
[5]Izaña Atmospheric Research Centre, (IARC/CIAI), AEMet, Santa Cruz de Tenerife, Spain.
[6]Department of Chemistry (T.U. Analytical Chemistry), Faculty of Science, University of La Laguna, Tenerife, Spain.
[7]Andalusian Institute for Earth System Research, IISTA-CEAMA, University of Granada, Granada, Spain.
[8]Applied Physics Department, University of Granada, Granada, Spain.

E-mail: elisabeth.alonso@ciemat.es

Abstract. The present work is a first approach to the study of the spatio-temporal variability of the submicrometer atmospheric aerosol in Spain. The aerosol measurements have been obtained simultaneously at seven monitoring stations that compose the REDMAAS network during two measurement campaigns corresponding to summer and winter seasons.
In both summer and winter periods those measurement stations with a direct influence of anthropogenic emissions recorded the highest concentrations of particle number. In the summer campaign, the average daily pattern of the aerosol size distribution in the traffic and background urban stations was conditioned by the traffic emissions and secondary aerosol formation through photochemical reactions (new particle formation events, NPF). However, the secondary aerosol had a higher contribution to the aerosol total number concentration in the rural background and high-altitude stations. In the winter campaign, in all sampling sites with the exception of Izaña station, the traffic and domestic activity emissions had a greater contribution than secondary aerosol formation on particle number total concentration.
New particle formation events were identified at all sites during the summer period, and at sites without direct influence of anthropogenic emissions during the winter campaign. Some aerosol shrinkage processes were also observed at the Madrid and El Arenosillo stations.

IOP Conf. Series: Earth and Environmental Science **28** (2015) 012001 doi:10.1088/1755-1315/28/1/012001

1. Introduction

Aerosol size distribution plays a crucial role on climate forcing and therefore the knowledge on its spatial and temporal variability has important implications in the understanding of climate impact.

Compilations of aerosol measurements with spatio-temporal dimension are scarce in the literature. Typically, these works are framed within aerosol instrumentation networks such as [1] focused on submicrometer atmospheric aerosol measurements in Germany obtained from the German Ultrafine Aerosol Network (GUAN, 2008) or [2] presenting submicrometer atmospheric aerosol measurements in Europe carried out within the Aerosols, Clouds, and Trace gases Research InfraStructure (ACTRIS) Network (www.actris.net).

The strong climate variability and geographical features diversity confer unique characteristics to the study of the atmospheric aerosol in Spain. Works that characterize the submicrometer atmospheric particles in different areas of the Spanish territory are numerous, as those developed by [3], [4], [5], [6], [7] and [8]. Nevertheless, to the moment there is no study on the variability of the characteristics of submicron aerosol size distributions in Spain on a temporal and spatial scale basis.

With this aim, measurements at different sites of the Spanish Network of Environmental DMAs (REDMAAS: http://www.redmaas.com) have been analysed. The aerosol measurements were simultaneously obtained at seven monitoring stations corresponding to the six research groups that form the REDMAAS. Two measurement periods, under different meteorological conditions, a summer period (1-30 June 2012) and a winter period (13 December 2012-15 January 2013) were selected for the study.

The main objective of this study was to characterize, for the first time in Spain, the influence of the meteorological conditions and the characteristics of the sampling sites on the features of submicrometer atmospheric aerosol size distributions.

2. Measurement campaigns

2.1. Measurement sites

The seven monitoring stations are located throughout the entire Spanish territory being representative of the climate and territorial variability of the country. A summary of the characteristics of each measurement station is shown in Table 1.

Table 1. Characteristics of the seven observation sites within Spain: research group/city, type of measurement station (Urban Traffic (UT), Urban Background (UB), Rural Background (RB) and High-Altitude (HA)), location (geographical coordinates and altitude) and climate.

GEOGRAPHICAL AREA AND CLIMATOLOGY		
Research group/City	**Measurement station type/Location**	**Climate**
IUMA-UDC: A Coruña	UB: 43.3 N, 8.4 W, 45 m a.s.l.	Atlantic
CIEMAT: Madrid	UB: 40.5 N, 3.7 W 657 m a.s.l.	Continental Mediterranean
IDÆA-CSIC: Barcelona	UB: 41.4 N, 2.1 E 68 m a.s.l.	Mediterranean
IDÆA-CSIC: Montseny	RB: 41.8 N, 2.4 E 720 m a.s.l.	Mediterranean
INTA: El Arenosillo (Huelva)	RB: 37.1 N, 6.7 W, 40 m a.s.l.	Ocean Mediterranean
CEAMA-UGR: Granada	UT: 37.2 N, 3.6 W, 680 m a.s.l.	Continental Mediterranean
IARC-AEMET: Izaña (Tenerife)	HA: 28.3 N, 16.5 W, 2367 m a.s.l.	Temperate Mountain

2.2. Measuring Instruments

During the campaigns, continuous measurements of aerosol size distributions have been performed with Scanning Mobility Particle Sizers (SMPS) in all sampling stations, with the exception of Granada station. For this site, an Ultrafine Particle Monitor (UFPM) was used.

SMPS equipments are composed by a Differential Mobility Analyser (DMA) connected to a Condensation Particle Counter (CPC), while UFPM is composed by a DMA and an aerosol electrometer used as detector. Detailed information on the equipments used in the aerosol monitoring campaigns can be found in table 2.

Table 2. Characteristics of the measurement equipment on each sampling station.

Equipment	Neutralizer/charger	Flow rate	Measurement size range/ Temporal resolution	Nº Channels (Decade)
IUMA-UDC: A Coruña				
TSI-SMPS: DMA 3081 CPC 3785	Kr-85 (2 mCi)	10:01	7.37/7.64-299.6/289 (4.5 min)	104 (64)/ 102 (64)
CIEMAT: Madrid				
TSI-SMPS: DMA 3081 CPC 3775	Kr-85 (2 mCi)	10:01	20.2-661.2 (4.5 min)	107 (64)
IDÆA-CSIC: Barcelona				
TSI-SMPS: DMA 3081 CPC 3772	Kr-85 (2 mCi)	5:01	10.9-478.3 (5 min)	106 (64)
IDÆA-CSIC: Montseny				
SMPS-IfT: DMA CPC 3772	Kr-85 (2 mCi)	5:01	10.9-478.3 (5 min)	71 (38)
INTA: El Arenosillo (Huelva)				
TSI-SMPS: DMA 3081 CPC 3776	Kr-85 (2 mCi)	10:01	14.1-736.5 (10 min)	111 (64)
CEAMA-UGR: Granada				
TSI- UFPM 3031	Corona-jet charger	5	20->200 (10 min)	5
IARC-AEMET: Izaña (Tenerife)				
TSI-SMPS: DMA 3081 CPC 3010	Kr-85 (10 mCi)	5:01	10.6-495.8 (5 min)	108 (64)

The instrument intercomparison within the REDMAAS network carried out prior to measurement campaigns [9, 10] guaranteed the quality of measurements.

In order to compare SMPS measurements, the data have been corrected for diffusion losses and multiple charges with the TSI AIM (Aerosol Instrument Manager, TSI Inc.) software. Furthermore, aerosol data obtained at Madrid, El Arenosillo and Izaña stations have also been corrected for particle diffusion losses in the entry sampling lines to the SMPS [11].

Total particle number concentration (N_t), and the number of particle concentrations for each of the three modes: nucleation ($N_{<30\ nm}$), Aitken ($N_{30\text{-}100\ nm}$) and accumulation ($N_{>100\ nm}$), were obtained from the aerosol size distributions. The choice of the intervals of particles size that define each of the modes is based on the classification developed by [12].

The average data coverage for each of the measurement campaigns (summer and winter periods) were 61% and 74% respectively. Figure 1 shows the data coverage for each measurement station and study period.

In addition, the local new particle formation (NPF) events have been identified based on the methodology developed by [12]. This methodology has also allowed the identification of shrinkage events.

For these events, the growth (GR) and shrinkage rates (SR) were calculated based on [13]. The calculation was made from the modes of the size distributions averaged every 15 or 10 minutes (D_{mode}), depending on the temporal resolution of each instrument (table 2). The aerosol size distributions were fitted to a lognormal function to estimate the mode.

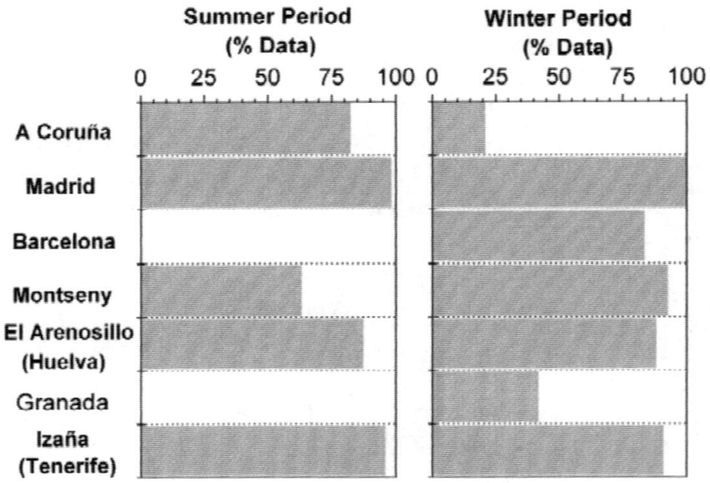

Figure 1. Data coverage of the submicron aerosol fraction for each measurement campaign and station.

3. Results and discussion

3.1. Particle number size distributions

The meteorological situation is rather different between both periods studied. In general, an active atmospheric dynamic together with an increase of the convective boundary layer facilitate the dilution processes in the warm period. However, in the cold period an increase of the anthropogenic emissions together with a greater frequency of high pressure systems over Iberian Peninsula lead to a high atmospheric stability and consequently to an increase of the atmospheric pollution load.

This situation caused differences in averaged total particle concentration between the two measurement periods in the different sampling sites.

With the exception of the Madrid station, data gaps from the traffic and background urban stations have made difficult the comparisons between the measurement periods (figure 1). In Madrid, N_t levels during the cold period were higher than during the warm period. This fact was due to the more stable conditions and reduction of the boundary layer along with the increase of pollutant emissions from traffic and domestic activities in this period [8]. On the opposite, at the rural background and high-altitude stations, with low or zero direct anthropogenic emissions, N_t was greater during the summer period because a greater atmospheric dynamics and the presence of gaseous precursors in the atmosphere facilitated the nucleation processes.

Furthermore, the emission sources in each sampling station conditioned the daily evolution of the aerosol size distribution in each measurement period.

At those stations with a clear influence of traffic emissions, such as A Coruña, Madrid, Barcelona and Granada three peaks were observed in the particle concentration in the summer period: the first peak (between 06:00 and 09:00 UTC approximately) and the third peak (between 18:00 and 00:00 UTC approximately) were attributed to traffic emissions, and the second peak (around 12:00 UTC) corresponded to secondary aerosol formation by photochemical activity. However, during the winter period only the two peaks corresponding to traffic emissions were observed. In these urban stations, the major contributor to total particle concentrations during the winter period was the Aitken mode, while, the increase in nucleation processes had a significant contribution to N_t, especially when NPF occurred, during summer period.

The situation was very different in those stations without a direct influence of anthropogenic emissions. The daily pattern of the aerosol size distribution was quite similar for both study periods. In the El Arenosillo and Montseny stations the particle concentration maximum appeared after 12:00

UTC. In these stations, the transport of the anthropogenic emissions from large urban areas imposed a background particle concentration throughout the day. This meant that the Aitken mode was the dominant particle mode, except during the central hours of the day when the environmental conditions were optimal for NPF development (low concentrations of preexisting particles in the atmosphere, high solar activity and low relative humidity), and consequently these processes were intense.

Only a peak was observed at the Izaña station around 12:00 UTC in the warm and cold periods, which was attributed to nucleation processes. In this station the nucleation mode was the largest contributor to the total particle concentration in both periods.

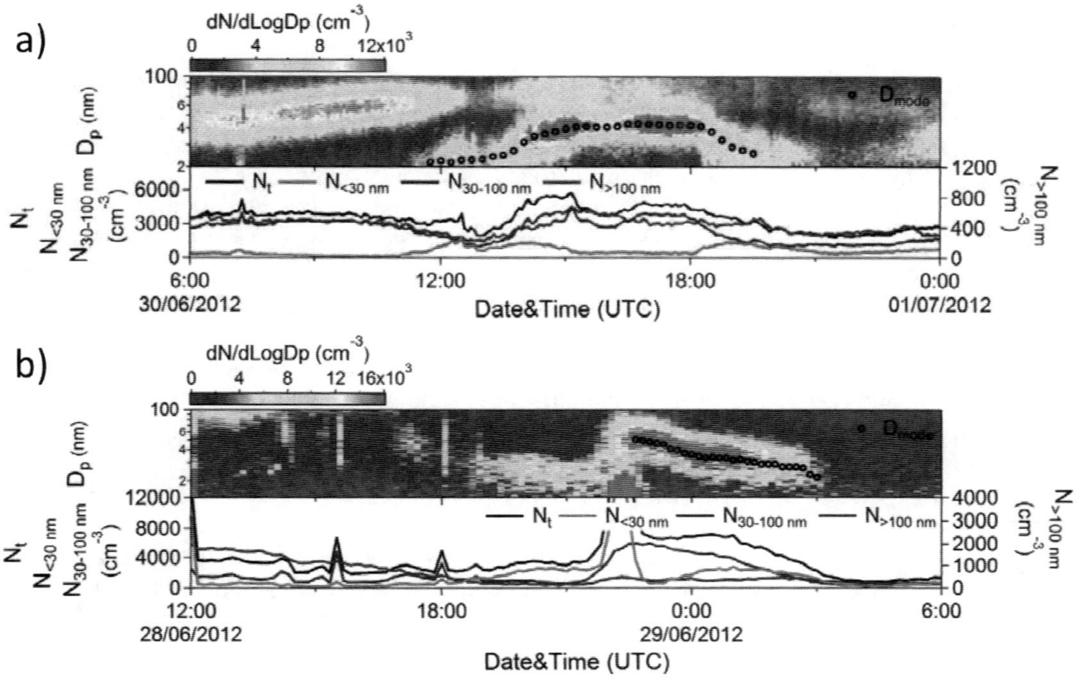

Figure 2. Example of two case studies: a) shrinkage during NPF identified in Madrid station and b) shrinkage event in absence of previous NPF identified in El Arenosillo station. Evolution of the aerosol size distributions, the total number of particles (N_t) and particle concentration for each of the three modes; nucleation ($N_{<30\ nm}$), Aitken ($N_{30\text{-}100\ nm}$) and accumulation ($N_{>100\ nm}$) as well as the particle mode diameter (D_{mode}) for each case study.

3.2. NPF and shrinkage events

NPF events have been observed during the summer period in all sampling stations, when the environmental conditions were optimal for their development.

These processes were particularly common in Izaña station associated with the transport of gaseous precursors from the boundary layer [4]. In Madrid, Montseny and El Arenosillo sites the influence of the biogenic volatile organic compounds (BVOCs) emitted by vegetated areas close to the stations played an important role in these processes [5, 7, 8]. In A Coruña and El Arenosillo the sea breezes had an important role in the NPF development [3, 6, 7].

In addition, these events were also identified during the winter period in Monseny, Izaña and El Arenosillo sites, stations without direct anthropogenic emissions.

The shrinkage events have been observed only during the summer period in Madrid and El Arenosillo stations. Two clear examples of these processes are shown in figure 2. In Madrid station a shrinkage process was identified during the growth phase of the newly nucleated particles on 30 June 2012. However, in El Arenosillo station the shrinkage was identified in the absence of a prior growth process on 28 June 2012. Both case studies occurred in the second half of the day, after 18:00 UTC, possibly caused by an increase in the wind speed and/or a reduction in the photochemical activity. The SR was -6.2 and -7.8 nm h^{-1} respectively, for each shrinkage process identified.

4. Conclusions

In each sampling site, the pollution sources and meteorological conditions determined the aerosol size distribution characteristics.

In both summer and winter periods, the urban area emissions were the main contributors to the total particle concentration at traffic and background urban stations, while in rural background and high-altitude stations, the nucleation processes had a higher contribution to aerosol particle number concentrations.

In general, the station location determined the behaviour of the particle concentration corresponding to each size distribution mode. At those stations with a higher influence of anthropogenic emissions the Aitken mode was the largest contributor to the total concentration of particles in opposition to less polluted sites, in which it was the nucleation mode.

NPF events occurred during the summer period in all measurement stations and also during the winter period in those with low anthropogenic influence. In addition, some shrinkage processes were identified in two measurement stations, Ciemat and El Arenosillo.

These results are a first approximation to a further study on the patterns of the submicrometer particle size distributions in Spain, which will be developed as part of the project "Atmospheric aerosol properties under different spatial and temporal scenarios and their climate influence (PROACLIM)". In addition, in a later study the influence of the geographic location, the land features and the atmospheric dynamics at regional level of each measurement site on aerosol size distribution and its processes will be examined in detail.

Acknowledgments

This work has been financed by the Ministry of Science and Innovation (CGL2011-15008-E, CGL2010-1777, CGL2011-27020, CGL2014-52877-R & CGL2014-55230-R), Xunta de Galicia (GRC2013-047 potentially cofounded by ERDF) and the European Union Seventh Framework Programme (FP7/2007–2013) ACTRIS under grant agreement no. 262254. E. Alonso acknowledges the FPI grant to carry out the doctoral thesis/PhD at the CIEMAT and M. I. García acknowledges the grant of the Canarian Agency for Research, Innovation and Information Society (ACIISI) co-funded by the European Social Funds.

References

[1] Birmili W *et al* 2009 *Gefahrstoffe Reinhaltung der Luft* **69** pp 137-145
[2] Asmi A *et al* 2011 *Atmos. Chem. Phys.* **11** pp 8893-8976
[3] Iglesias-Samitier S *et al* A14 *Proc. of 2nd Iberian Meeting on Aerosol Science and Technology (RICTA) Tarragona, Spain, 7–9 July 2014* pp 79
[4] García M I *et al* 2014 *Atmos. Chem. Phys.* **14(8)** pp 3865-3881
[5] Cusack M *et al* 2013 *Tellus B* **65** pp 1–19
[6] Sorribas M *et al* 2011 *Atmos. Chem. Phys.* **11(21)** pp 11185-11206
[7] Sorribas M *et al* 2015 *Science of the Total Environment* **511** pp 723-737
[8] Gómez-Moreno F J *et al* 2011 *Atmos. Environ.* **45** pp 3169-3180
[9] Gómez-Moreno F J *et al* B-WG01S2P30 *Proc. of European Aerosol Conference (EAC), Granada, Spain, 2–7 Sept. 2012*
[10] Gómez-Moreno F J *et al* A076 *Proc. of European Aerosol Conference (EAC), Prague, Czech Republic, 1–6 Sept. 2013*
[11] Wiedensohler A *et al* 2012 *Atmos. Meas. Tech.* **5** pp 657-685
[12] Charron A and Harrison R M 2013 *Atmos. Environ.* **37** pp 4109-4119
[13] Dal Maso M *et al* 2005 *Boreal Env. Res.* **10** pp 323
[15] Kulmala M *et al* 2012 *Nature protocols* **7** pp 1651-1667

IOP Conf. Series: Earth and Environmental Science **28** (2015) 012002 doi:10.1088/1755-1315/28/1/012002

Hygroscopic Properties of Atmospheric Aerosol Measured with an HTDMA in an Urban Background Site in Madrid

E Alonso-Blanco, F J Gómez-Moreno, M Becerril, E Coz and B Artíñano

Department of Environment, CIEMAT, Madrid, Spain

E-mail: elisabeth.alonso@ciemat.es

Abstract. The observation of high aerosol hygroscopic growth in Madrid is mainly limited to specific atmospheric conditions, such as local stagnation episodes, which take place in winter time. One of these episodes was identified in December 2014 and the hygroscopic growth factor (GF) measurements obtained in such episode were analysed in order to know the influence of the meteorological conditions on aerosol hygroscopic properties.

The prevailing high atmospheric stability triggered an increase of the particle total concentration during the study period, with several peaks that exceeded 4.0 10^4 particles cm^{-3}, as well as an increase in the inorganic fraction of the aerosol, the NO_3^- concentration, which in this case corresponded to 25% of the total PM_1 non-refractory composition.

The aerosol hygroscopic growth distribution was bimodal during the episode, with an average GF around 1.2 for the five dry particle sizes measured and an average GF spread $\geq$ 0.15. In addition, it is important to note that when a reduction in the concentrations of NO_3^- is observed, it coincides with a decrease of the GF and its spread. These data suggest, on the one hand, a high degree of external mixing state of the aerosol during the episode and, on the other hand, a notable association between the GF and the inorganic fraction of the aerosol.

1. Introduction

Hygroscopicity is one of the most uncertain properties of the atmospheric aerosol [1, 2], principally because it involves two aspects: the aerosol composition and size, and the humidity/temperature ambient conditions. The growth that atmospheric aerosols may experience due to their hygroscopicity is relevant in some processes in which they participate, as the direct and indirect aerosol effects on radiative forcing and climate, in the atmosphere, or the potential damage to the health inside the human body.

The hygroscopic behaviour of atmospheric aerosols based on measurements of the particle size through a Hygroscopic Tandem Differential Mobility Analyzer (HTDMA) has been investigated at different European locations, especially in Northern Europe, e.g. [3], [4] or [5]. However, such investigations are scarce in Southern Europe [6, 7], with rather different climatic conditions.

In the city of Madrid, the primary sources of pollutants determining the aerosol composition, and consequently its hygroscopicity, are emissions from traffic and domestic activities, predominantly of carbonaceous nature. As it will be seen in this study, the meteorological conditions of this geographical area will also play an important role on the aerosol hygroscopic properties.

This work analyzes the aerosol growth during a pollution episode in the 2014 winter season. To this aim, the aerosol GF measurements have been complemented with data on aerosol size distribution, aerosol composition and meteorological conditions.

2. Sampling site and measurements

This study has been conducted during winter 2014 in an urban background area located at CIEMAT facilities (40º 27′ 23.2″ N, 03° 43′ 32.3″ E and 650-700 m asl), in Madrid.

During this season local stagnation episodes are usual. The meteorological scenario corresponds to an anticyclone situation over the Iberian Peninsula, leading to a high atmospheric stability and consequently to the presence of intense thermal inversions, which are constant throughout this type of episodes [8, 9, 10].

A wide variety of instruments have been measuring the different properties of atmospheric aerosols at the CIEMAT station complemented with meteorological parameters:

1) The hygroscopic growth factor of atmospheric aerosol has been obtained with an HTDMA. The measurements have been acquired at RH of 90% for five dry particle sizes (50, 80, 110, 190 and 265 nm) covering the submicrometer fraction of the atmospheric aerosol. The temporal resolution of the measurements was 3 min. More detailed information on this equipment can be found in [11] and [12].

The measurements have been corrected by the $TDMA_{inv}$ program [13] and the GF has been recalculated at 90% RH from $TDMA_{inv}$ for RH variations of ±2% about target value (90% RH). After correcting the RH, data coverage of 90 % for the study period was reached. The number of lost scans was higher when the dry particle size selected was increased.

The GF values collected in this work corresponded to GF_{avg} provided by the $TDMA_{inv}$. This parameter was defined as the growth factor that would be observed if the absorbed water were equally distributed among all particles in a sample [3]. In addition, the spread of growth factor was used to evaluate the aerosol mixing state; a GF spread ≤ 0.10 indicates an internal mixture, or a quasi-internal mixture, while a GF spread ≥ 0.15 can be considered externally mixed, or quasi-internally mixed [3, 4].

2) The aerosol size distribution between 14 and 660 nm has been provided by a Scanning Mobility Particle Sizer (TSI-SMPS: DMA3081 with CPC 3775), with temporal resolution of 4.5 min. The particle concentrations for each of the three modes, nucleation ($N_{<30\ nm}$), Aitken ($N_{30\text{-}100\ nm}$) and accumulation ($N_{>100\ nm}$), were obtained from the aerosol size distributions.

3) The PM_1 non-refractory composition (organics, NO_3^-, SO_4^{2-}, NH^{4+} and Cl^-) has been obtained by an Aerosol Chemical Speciation Monitor (ACSM, Aerodyne Research, Inc.). The ACSM resolution was around 30 min for each species.

4) A meteorological tower registered data on precipitation, pressure, temperature at two levels (obtained with a difference of 50 m), relative humidity (RH) and wind speed and direction at the measurement site every 10 minutes.

Additional information on PM_{10}, $PM_{2.5}$, NO_2 and O_3 was provided by the suburban station of Casa de Campo (3° 44' 50.44" W, 40º 25' 09.68" N, 645 m asl) located at 4.5 km South from the measuring place. This station belongs to the Air Quality Monitoring Network of the Madrid municipality and is representative of the environmental characteristics of the study area.

Furthermore, the pressure field maps at surface level provided by UK Meteorological Service (http://www.metoffice.gov.uk/) have been used to interpret the atmospheric meteorological situation at synoptic scale.

3. Results and discussion

3.1. Identification of the episode

A local stagnation episode was identified from 14 to 28 December 2014. The synoptic maps showed a high pressure system over the Iberian Peninsula, which remained during the whole episode inhibiting the dispersion of pollutants (Figure 1).

The average temperature for the episode was 7.1±1.0°C, with the presence of intense thermal inversions during night time. In addition, precipitation was not observed and the average RH was 74±9%.

The average wind speed was 2.7±1.5 m/s and the wind direction in the area followed the typical pattern with a NE-SW steering axle, characteristic of the measurement area [14].

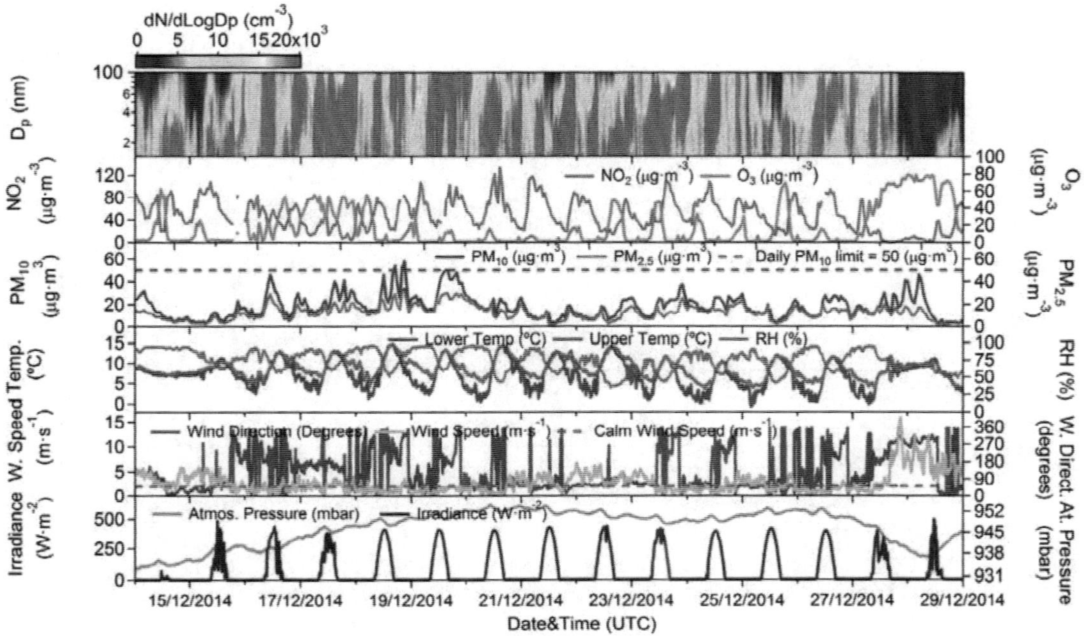

Figure 1. Evolution of the aerosol size distribution and meteorological conditions (temperature, relative humidity, wind speed, wind direction, and irradiance) obtained in the Ciemat station as well as NO_2, O_3, PM_{10}, and $PM_{2.5}$ concentrations provided by the air quality station Casa de Campo during the local stagnation episode.

The evolution of the aerosol size distribution indicated a high particle concentration throughout the period, which as evidenced by the high concentrations of NO_2, possibly originated from traffic sources. Nevertheless, the daily PM_{10}limit value (24 hours) established by Spanish legislation (*Royal Decree* **102/2011** of January 28: 50 $\mu g \cdot m^{-3}$), based on the *European Directive* **2003/17/CE**, was not exceeded. However, PM_{10} concentrations higher than 50 $\mu g \cdot m^{-3}$ were observed for some time intervals during the study period. Also high $PM_{2.5}$ concentrations higher than 25 $\mu g \cdot m^{-3}$ were also recorded.

3.2. Aerosol growth factor during the local stagnation episode

The characteristics of the meteorological scenario indicated in the previous section not only triggered an increase in the pollutant concentration but also their chemical aging. This facilitated an increase in the inorganic fraction of atmospheric aerosol.

The total particle concentration was high throughout the period, with an average value of $16 \times 10^3 \pm 1 \times 10^3$ particles·cm^{-3} and peaks that exceeded 40×10^3 particles·cm^{-3} (Figure 2). The PM_1 non-refractory composition during the local stagnation episode was mainly organic, 62% of the total composition, followed by NO_3^- (25%).

During the episode, the aerosol growth factor distribution was bimodal for all measurements, with an average GF centred on 1.2 for the five dry particle sizes. In addition, on average for the study period, the GF spread for the particle sizes measured was ≥ 0.15, which suggested a high degree of external mixing of the aerosol. It has to be emphasized that the GF spread was high for larger particles i.e. for a dry size of 50 nm it was 0.15±0.02 while for a dry size of 265 nm was 0.19±0.07.

IOP Conf. Series: Earth and Environmental Science **28** (2015) 012002 doi:10.1088/1755-1315/28/1/012002

In relation to the GF daily variation, the GF maximum corresponded to a maximum concentration of NO_3^- as shown in figure 2. Furthermore, the situation observed on 21 and 22 December 2014 should be noted. Wind speed was greater than 3 $m \cdot s^{-1}$ and NO_3^- concentration did not rise above 0.3 $\mu g \cdot m^{-3}$ on both days. This involved a significant reduction of the GF in the five particle sizes measured, which did not reach 1.2 (average value during the study period). This scenario was also observed in the GF spread, which indicated a reduction of the mixing state of aerosol in all the measured particle sizes.

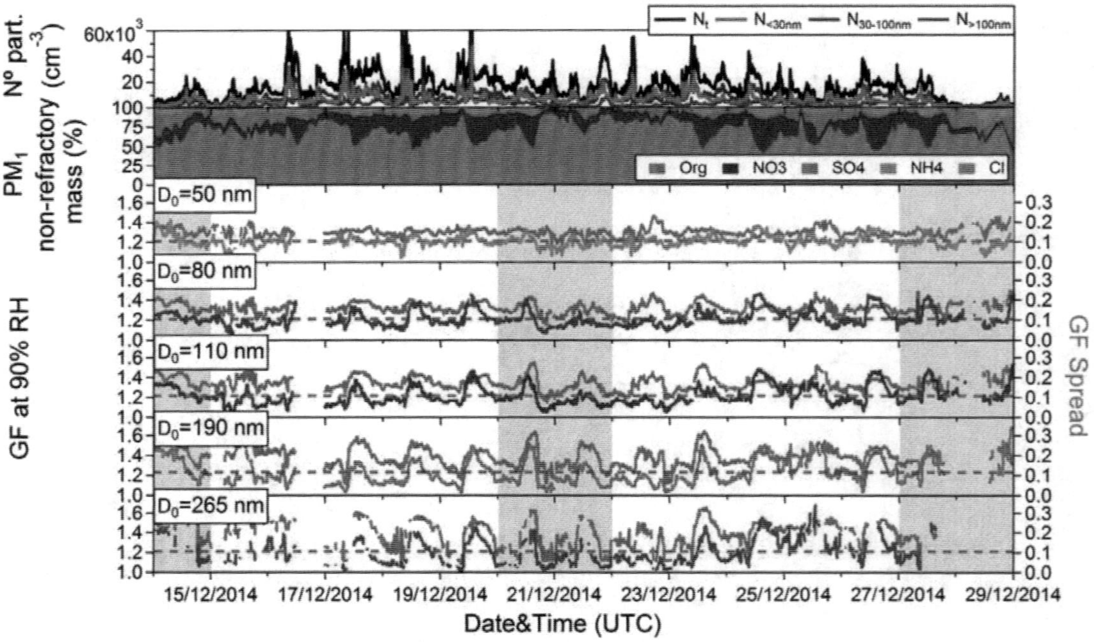

Figure 2. Temporal evolution of GF and its spread for dry particle sizes (D_0) of 50, 80, 110, 190, 265 nm at RH=90% as well as the evolution of the PM_1 non-refractory composition and modal particle concentrations during the local stagnation episode. The red line on each plot represents the average GF for each measured particle size. For a better interpretation of pollutant data, the weekend period has been indicated by a shaded area.

4. Conclusions

The aerosol hygroscopicity has been studied during a winter episode under local stagnation conditions in Madrid.

The strong atmospheric stability during the episode influenced the chemical composition of the atmospheric aerosol and consequently its hygroscopicity.

A high concentration of particles was observed, exceeding 40×10^3 particles$\cdot cm^{-3}$ during several time intervals. Also, the formation of ammonium nitrate was high in such events, reaching 25% of the PM_1 non-refractory composition.

The aerosol growth distribution measurements were bimodal for the entire episode, with an average GF of 1.2 for the five dry particle sizes measured and a GF spread ≥ 0.15. In those days, when the formation of NO_3^- was reduced, the GF and its spread also decreased. This situation suggested that the increase of the inorganic fraction of the aerosol during the episode led to a high degree of aerosol external mixing in all measured particle sizes.

These preliminary results will be part of a deeper study which will compare the aerosol hygroscopic growth in Madrid under different meteorological scenarios.

Acknowledgments
This work has been supported by the Spanish Ministry of Science and Innovation funding the MICROSOL (CGL2011-27020) and PROACLIM (CGL2014-52877R) projects and Ramón Areces Foundation, the AEROCLIMA project (CIVP16A1811) and Autonomous Community of Madrid, the TECNAIRE project (P2013/MAE-2972). The authors are grateful to Martin Gysel for the development of $TDMA_{fit}$ algorithm to invert HTDMA data ($TDMA_{inv}$) and allowing free use within the scientific community.
E. Alonso-Blanco acknowledges the FPI grant to carry out the doctoral thesis/PhD at the Research Center for Energy, Environment and Technology and of Economy and M. Becerril the FPI predoctoral research grant BES-2012-056545 at the Spanish Ministry of Economy and Competitiveness (MINECO).

References

[1] IPCC 2013 *Working Group I Contribution to the IPCC Fifth Assessment Report Climate Change*
[2] Varghese S K and Gangamma S 2009 *Aerosol Air Qual. Res.* **21(7)** pp 360-379
[3] Holmgren H *et al* 2014 *Atmos. Chem. Phys.* **14(18)** pp 9537-9554
[4] Sjogren S *et al* 2008 *Atmos. Chem. Phys.* **8** pp 5715-5729
[5] Massling A *et al* 2005 *Atmos. Environ.* **39** pp 3911-3922
[6] Adam M *et al* 2012 *Atmos. Chem. Phys.* **12(13)** pp 5703-5717
[7] Bezantakos S *et al* B-WG01S2P50 *Proc. of European Aerosol Conference (EAC), Granada, Spain, 2–7 Sept. 2012.*
[8] Crespi S N *et al* 1995 *J. App. Meteo.* **34(7)** pp 1666-1677
[9] Artíñano B. *et al* 2003 *Environ. Poll.* **125** pp 453-465
[10] Coz E *et al* 2010 *J. Geophys. Res.* **115** D12204
[11] Alonso-Blanco E *et al* A06 *Proc. of 2nd Iberian Meeting on Aerosol Science and Technology (RICTA), Tarragona, Spain, 7–9 July 2014* pp 33
[12] Nilsson E. *et al* 2009 *Atmos. Meas. Tech.* **2** pp 313-318
[13] Gysel M *et al* 2009 *J. Aero. Sci.* **40(2)** pp 134-151
[14] Salvador P 2014 *Doctoral Thesis. Faculty of Physics, University Complutense of Madrid, Madrid (Spain)*

IOP Conf. Series: Earth and Environmental Science **28** (2015) 012003 doi:10.1088/1755-1315/28/1/012003

Particulate matter formation from photochemical degradation of organophosphorus pesticides

E Borrás, M Ródenas, T Vera and A Muñoz

Fundación CEAM. Parque Tecnológico, Paterna, Valencia, Spain.

E-mail: esther@ceam.es

Abstract. Several experiments were performed in the European Photo-reactor – EUPHORE – for studying aerosol formation from organophosphorus pesticides such as diazinon, chlorpyrifos, chlorpyrifos-methyl and pirimiphos-methyl. The mass concentration yields obtained (Y) were in the range 5 – 44 % for the photo-oxidation reactions in the presence and the absence of NOx. These results confirm the importance of studying pesticides as significant precursors of atmospheric particulate matter due to the serious risks associated to them.

The studies based on the use of EUPHORE photoreactor provide useful data about atmospheric degradation processes of organophosphorus pesticides to the atmosphere. Knowledge of the specific degradation products, including the formation of secondary particulate matter, could complete the assessment of their potential impact, since the formation of those degradation products plays a significant role in the atmospheric chemistry, global climate change, radiative force, and are related to health effects.

1. Introduction

Pesticides are most widely used chemical compounds and actually, they are emerging chemical pollutants precursors of particulate matter. Today's agriculture involves heavy use of synthetic pesticides, including herbicides, fungicides, acaricides, insecticides and avicides. Their entry into the atmosphere occurs during application or subsequent processes such as volatilization and re-suspension [1]. The number of patented pesticides is very high and many families are authorized, prohibited or banned by regulations – e.g., Regulation (EC) 1107/2009 – due to the restricted regulations about human and environmental safety. As consequence, the scientific knowledge about the behaviour of pesticides in the atmosphere is highly demanded due to their use in Mediterranean agriculture in a broad range of applications. However, there is a relative lack in the scientific knowledge about the behaviour of pesticides in the atmosphere.

Organophosphorus pesticides are extensively used in agriculture, horticulture, and a variety of household applications. They are chemical or biological products used to prevent diseases, kill, repel, control plagues or interrupt the growth of plants, including herbicides, acaricides, and insecticides. Once pesticides are applied in the field, it can be partitioned into the soil, water, and atmosphere and, consequently, they have a significant environmental impact. Pesticides can be emitted into the atmosphere through dispersion during spraying or volatilization from ground or leaf surfaces, and the amount emitted is a function of their physical properties and their manner of application. Associate to their low-volatile or semi-volatile nature, they have a distribution into gas and condensed phases more or less displaced depending on their physicochemical properties and environmental conditions [2]. Around a hundred insecticides with the organophosphorus structure can be classified in chemical families showing a wide range of efficacies and action modes. An important group include esters, amides, or thiol

derivatives from phosphoric acid (R1OP(=O)(OR2)(OR3)) or from phosphorothioic acid (R1OP(=S)(OR2)(OR3)). Some examples of the most used organothiophosphorus pesticides in Europe are chlorpyrifos, chlorpyrifos-methyl, diazinon, and pirimiphos-methyl among others.
The major routes of degradation of pesticides in the troposphere involve photolysis, ozonolysis and reactions with hydroxyl and nitrate radicals. The result is a reduction of their concentration in air. However particulate matter (PM) is formed and it could have a toxicity, residence time and atmospheric chemical stability different than the original molecule. Relevant information for several of the most Mediterranean area world-wide used organophosphorus insecticides has been obtained in a high volume atmospheric simulation chambers – EUPHORE - with an extensive instrumentation for monitoring the particulate matter formed.

2. Experimental Section

2.1 Reactor and instruments
The experiments were carried out in the high volume outdoor smog chambers EUPHORE (EUropean PHOtoREactor) (Valencia, Spain). These simulator photoreactors were designed to work under near-realistic atmospheric conditions minimizing losses and wall-interactions effects. These chambers consist of two half spherical fluoropolymeric bags, each one of 200 m3 with integrated measuring systems for monitoring pressure, humidity, temperature, precursor species, and reaction products [3]. Pressure, humidity and temperature were measured using a pressure sensor (Air-DB-VOC, Sirsa, Madrid, Spain) and a dew point hydrometer (TS-2, Walz, Effeltrich, Germany). An Eco Physics AG (AL-ppt-77312, Duernten, Switzerland), an API NO_x monitor (API200AU, Teledyne API, San Diego, USA) and an NOx analyzer (ML9841A, Teledyne Monitor, Englewood, USA) were used for measuring NO, NO_2 and NO_x. A White-type mirror system (path length of 553.5 m) coupled to a Fourier Transform Infrared spectrometer with MCT detector (NICOLET Magna 550, Thermo Scientific, USA) was used for recording concentrations of selected pesticide, nitrous acid (HONO) and SO_2. Retrieval of concentration profiles were calculated by using specific software [4]. FTIR calibration procedure consisted on several introductions of each quantified compound, using standards, into the EUPHORE simulation chamber at known concentrations. Aerosol mass concentration was measured with two on-line instruments. One was a scanning mobility particle sizer (SMPS), model 3080 (TSI, Shoreview, USA). This system measured size distributions in the 11-789 nm diameter range in real time with a 5 min scan rate, and it provides aerosol concentrations assuming spherical shapes and multi-charge correction for the condensed organic material. Sheath and aerosol sampling flows were 3 L min^{-1} and 0.30 L min^{-1}, respectively. The other automated instrument was a tapered element oscillating monitor (TEOM) (model 1400a, Ruppercht and Patashnick, Albany, USA) with a 1 min scan rate and a sampling flow of 3 L min^{-1}.

2.2 Experiments
A blank chamber experiment, described in Borrás and Tortajada-Genaro, 2012, was performed to assure the absence of possible simulator artifacts due to the off-gassing of compounds from reactor walls. The photoreactor was filled with air from a purification system. Non-detectable hydrocarbons and nitrogen oxides were measured and the aerosol background was 0.010 ± 0.005 μg m^{-3} (60 part cm^{-3}). Specific experiments were performed for guaranteeing the correct injection of chlorpyrifos-methyl into the reaction chamber, because generally low-vapor pressure hydrophobic compounds present important problems (e.g. electrostatic effects with connections). Pesticide was injected via heated air stream (flow 10 L min^{-1}, temperature 350 ºC) through a short PTFE tube connection. Under these conditions, losses or decomposition processes were negligible before photochemical degradation reactions started (by opening doom covers).

The oxidation experiments consisted in the sun exposition under dry conditions (<2% RH, 295-298 K) in the absence of initial inorganic seeds and low concentrations of nitrogen oxides (Table 1). For photo-oxidation activated by NOx, HONO was generated by a liquid-phase

IOP Conf. Series: Earth and Environmental Science **28** (2015) 012003 doi:10.1088/1755-1315/28/1/012003

reaction between a 0.5% $NaNO_2$ solution and a 30% H_2SO_4 solution and transferred via a stream of purified air. Later, individual pesticides were introduced into the chamber in the range of 50 – 100 ppbV. After the reactants were mixed for 10 min, the chamber was exposed to natural sunlight ($JNO_2 \approx 6.8\times10^{-3}$ s^{-1}) and the reaction started. The onset of aerosol formation was considered to occur when the first significant particle concentration was registered (signal > 3σ background). The specific dilution process was determined using SF_6 as a tracer (2.3×10^{-5} s^{-1}). Size-resolved wall-loss corrections were not needed for aerosol measurements, since negligible variations were observed during filter sampling for particle number concentration. Data analysis was performed with the statistical package SPSS for Windows v 16.0.

3. Results

A detectable amount of particulate matter was obtained in all experiments and the aerosol yield (Y), or the capacity of organophosphorus pesticide to produce particles, was calculated from the equation proposed by Odum et al. (1996).

$$Y = \frac{M_0}{\Delta HC} \quad [1]$$

where *Mo* (µg m-3) is the aerosol mass concentration formed and *ΔHC* (µg m^{-3}) is the mass concentration of pesticide reacted. Yields, reported in Table 1, were thus calculated using the precursor concentration from FTIR data and the aerosol concentration from aerosol wall losses corrected SMPS data between the start and the maximum of aerosol formation.

Table1. Summary of aerosol yields from the different photo-oxidation conditions of organophosphorus selected pesticides.

Pesticide	Type of Experiment	Y (%)
Diazinon	Photolysis	< 0.1
	Photo-oxidation in the presence of NOx	44
Chlorpyrifos	Ozonolysis	< 0.1
	Photolysis	6
	Photo-oxidation in the absence of NOx	-
	Photo-oxidation in the presence of NOx	23
Chlorpyrifos-methyl	Ozonolysis	< 0,1
	Photolysis	< 0,1
	Photo-oxidation in the absence of NOx	8,0
	Photo-oxidation in the presence of NOx	21,0
Pirimiphos-methyl	Ozonolysis	< 0,1
	Photolysis	< 0,1
	Photo-oxidation in the absence of NOx	5
	Photo-oxidation in the presence of NOx	15

The Y-values ranged between 2.1± 0.1 % and 44 ± 0.3 %. The results, obtained in the absence of aerosol seeds and temperatures ranged between 14.5 and 15.6 °C indicated that the atmospheric degradation of organophosphorus pesticides leaded to a reduced formation of products with a low vapor pressure (particulate products). Nevertheless, a higher aerosol formation is expected (2 or 3 folds) in the atmosphere, associated to an early gas-condensed phase partitioning of semi-volatile products in the ambient particles [5] or under lower ambient temperatures. Compared to other organic aromatic pesticides, the aerosol yields obtained under photo-oxidation conditions were similar to propachlor (15%), a chloride-substituted aromatic ring [6], and hymexazol (4.8%), a N-heterocycle compound [7] employing higher pesticide and NOx concentrations, in both cases. The aerosol yields were confirmed by calculating the curves of the aerosol mass concentration (Mo) as a function of the pesticide reacted after the onset aerosol formation (ΔHC). A strong linear correlation (R^2 average > 0.95) and statistical test (t-test) confirmed that calculated Y-values from both methods were comparable (p-value<0.05), confirming that the aerosol formation is a NO_x-dependent process.

4. Atmospheric implications

The extensive use of organophosphorus pesticides presents an important environmental effect, and there is a concern about the subsequent changes of chemical atmosphere dynamics and their toxicological hazards. In recent years, many studies have demonstrated that organophosphorus pesticides and some of their degradation products are mutagenic, carcinogenic, cytotoxic, genotoxic, teratogenic, and immunotoxic [8]. The present study, contributes providing useful data about diurnal pesticide degradation. Knowledge of the specific degradation products, including the formation of secondary particulate matter, could complete the assessment of their potential impact. So, the results have indicated that these pollutants are a relevant source of particulate matter. The formation of PM is important because they play a significant role of atmospheric chemistry since they are related to health effects, climate change and radiative force [9]. Hence, the results of this study can contribute to the selection of environmentally sustainable strategies against plagues. Thus, the understanding of atmosphere reactions should help to estimate the expected formation of gas and/or particulate products in the troposphere for each pesticide. Just then, the design and selection between related molecules will be correctly performed based on criteria of efficient action and low environmental impact of the pesticide and their residues.

Acknowledgments

The research leading to these results received funding from the Spanish Ministry of Economy and Competitivity (project IMPLACAVELES: CGL2013-49093-C2-1-R), and Generalitat Valenciana for the DESETRES-Prometeo II project.

References

[1] Gil Y and Sinfort C 2005 *Atmos. Environ.* **39** 5183
[2] SANCO/3061/99 – rev. 1.6. 2005
[3] Borrás E, Tortajada-Genaro L.A 2012a *Atmos. Environ.* **47** 154
[4] Ródenas M. Report for ESF-INTROP exchange grants, 2008 http://www.ceam.es/GVAceam/archivos/MRodenasINTROPreport.pdf.
[5] Ng NL, Kroll JH, Chan AWH, Chhabra PS, Flagan RC, Seinfeld JH 2007 *Atmos. Chem. Phys.* **7** 3922
[6] Muñoz A, Vera T, Sidebottom H, Ródenas M, Borrás E, Vázquez M, Raro M, Mellouki A, 2012 *Atmos. Environ.* **49** 40
[7] Tortajada-Genaro LA, Borrás E, Muñoz A 2013 *Chemosphere* **92** 1041
[8] Wang X, Mu Z, Shangguan F, Liu R, Pu Y, Yin L 2013 *www.plosone.org* **8** e66703
[9] Ramanathan V, Crutzen PJ, Kiehl JT, Rosenfeld D 2001 *Science* **294** 2124

IOP Conf. Series: Earth and Environmental Science **28** (2015) 012004 doi:10.1088/1755-1315/28/1/012004

Use of high-volume outdoor smog chamber photo-reactors for studying physical and chemical atmospheric aerosol formation and composition

E Borrás, M Ródenas, T Vera and A Muñoz

Fundación CEAM. Parque Tecnológico, Paterna, Valencia, Spain.

E-mail: esther@ceam.es

Abstract. The atmospheric particulate matter has a large impact on climate, biosphere behaviour and human health. Its study is complex because of large number of species are present at low concentrations and the continuous time evolution, being not easily separable from meteorology, and transport processes. Closed systems have been proposed by isolating specific reactions, pollutants or products and controlling the oxidizing environment. High volume simulation chambers, such as EUropean PHOtoREactor (EUPHORE), are an essential tool used to simulate atmospheric photochemical reactions. This communication describes the last results about the reactivity of prominent atmospheric pollutants and the subsequent particulate matter formation. Specific experiments focused on organic aerosols have been developed at the EUPHORE photo-reactor. The use of on-line instrumentation, supported by off-line techniques, has provided well-defined reaction profiles, physical properties, and up to 300 different species are determined in particulate matter. The application fields include the degradation of anthropogenic and biogenic pollutants, and pesticides under several atmospheric conditions, studying their contribution on the formation of secondary organic aerosols (SOA). The studies performed at the EUPHORE have improved the mechanistic studies of atmospheric degradation processes and the knowledge about the chemical and physical properties of atmospheric particulate matter formed during these processes.

1. Introduction

Environmental air pollution is generally classified as photochemical smog pollution, biomass burning, coal combustion and other, according to the source of contamination [1]. Photochemical smog is a mixture of primary and secondary pollutants that are formed by the irradiation of primary pollutants with UV light (290–400 nm). Indeed, gaseous species lead to a variety of secondary oxidized products – ozone (O3), aldehydes, ketones, carboxylic acids, peroxyacetyl nitrates, organic aerosols, etc... –, many of them being often more harmful than their precursors. Furthermore, atmospheric organic matter undergoes progressive oxidation during its chemical transformations, to yield products, generally oxygenated, that have sufficiently low vapour pressures to partition themselves between gas and aerosol phases [2]. These phase change processes affect the atmospheric aerosol by nucleation or by condensation of particles. Moreover, chemical transformations can also occur in the condensed phase. All these processes – in gas and in particulate phase – are susceptible to modify the physicochemical properties in a short time and affected by different conditions.

Since, in the ambient, it is difficult to separate the chemistry from meteorology, advection, dispersion, and other transport processes. Then, closed systems have been developed by isolating specific compounds of interest and controlling the oxidizing environment. The smog chamber facility is an essential tool used to simulate atmospheric photochemical reaction processes by irradiating primary pollutants and oxidants inside the chamber. Originally, smog chamber experiments were essentially focused on the understanding and developing of atmospheric oxidation mechanisms in the gas phase. Later, these facilities were also used to investigate processes concerning secondary organic aerosol (SOA) formation as well as concerning the physicochemical properties of aerosols [3].

However, environmental chambers are not without uncertainties in characterization and background effects depending of volume, light source, design, control of temperature, filling air system, etc [4]. In consequence, background chemicals, off-gassing materials and memory effects may complicate the chemical composition of the studied mixture and hence the reliability of the data [5]. This paper describes the design, analytical instrumentation and environmental applications of EUPHORE smog chamber developed to minimize reactor effects in studies of VOC reactivity, to increase knowledge about the chemical and physical characterization of degradation products in particulate phase. EUPHOPE is a large a Teflon – fluoropolymer - facility designed to carry out experiments under a great S/V ratio, realistic natural sunlight irradiation and purification developed air system. EUPHORE has been be used for studying degradation studies of aromatic hydrocarbons [2], biogenic compounds [6], mixture of pollutants [7], pesticides [8], such as an intercomparison platform, etc…that made these facilities ones of the most important and representative in the scientific community.

2. Experimental Section

2.1 Facility Description

Each simulation chamber (there are two identical smog chambers) consists of a half-spherical Teflon bag with a volume of 200 m^3, a diameter of 7 m, and a surface-to-volume ratio of 1 m^{-1}. The chambers were made of a fluoropolymer foil that has a transmission up to 80% of the solar radiation, in the wavelength of 280 – 640 nm. To compensate heating of the outdoor chamber by solar radiation, the chamber floor panels are cooled by a home-made refrigeration system. For mixing the air, two larges fans are installed at the bottom of the chamber with an air throughput of 4000 m^3 h^{-1} that permits an homogenous air distribution in height as well as in area. Moreover, each chamber is filled with purified air from an air purification system which included absorption driers (type HEA 1400, Zander, Essen, Germany) with a molecular sieve (ECO 30%, MOL 70% Sogimair, Barcelona, Spain), providing air that had non-detectable hydrocarbons, NO_x and particles (aerosol background of 0.010 ± 0.005 μg m^{-3}.

IOP Conf. Series: Earth and Environmental Science **28** (2015) 012004 doi:10.1088/1755-1315/28/1/012004

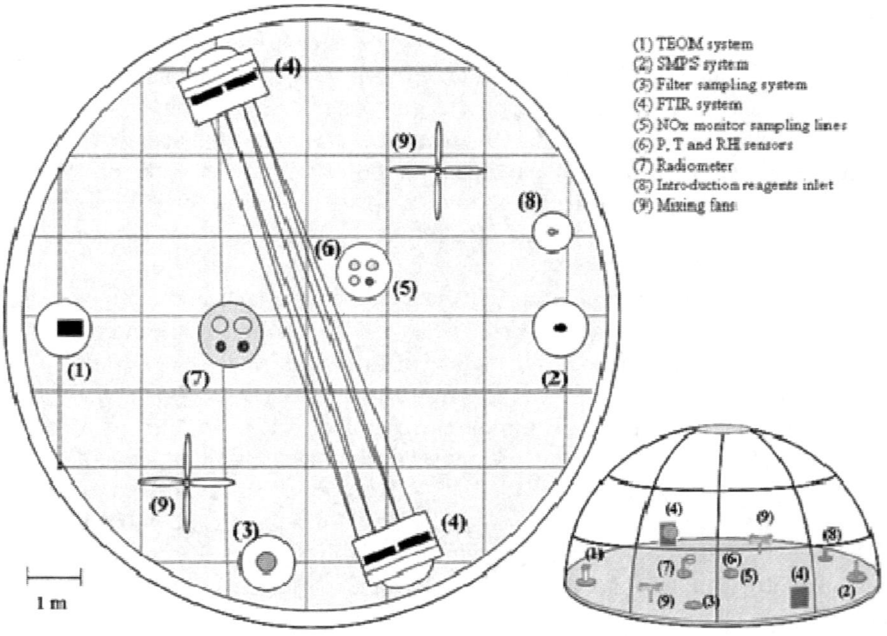

Figure 1. The schematic of the EUPHORE smog chamber facilities.

2.2 Aerosol Instrumentation

Aerosol Generator. The aerosol atomizer (TSI Model 9306) generates submicron aerosols. The mean particle size of the generated aerosol can be varied between 0.02 μm and 0.3 μm by atomizing a solution. More usually solutions employed are NaCl, $(NH_4)_2SO_4$, etc, at different concentration levels to generate different mass concentrations into the chamber. Combined with the aerosol generator, an aerosol neutralizer of Kr has been installed to reduce the number of charged particles generated.

Tapered element oscillating monitor (TEOM).This instrument model 1400a, was supplied by Ruppercht and Patashnick Co. Inc, (Albany, NY, USA). It is employed to measure real-time aerosol mass concentrations with a 1 min scan rate and PM1 sampling head. The particles are collected on a replaceable 0.5 cm diameter filter, incorporated in a microbalance that is free to oscillate. This sensor causes a flow of air, regulated by a thermal mass flow controller, through a filter placed in a holder which vibrates at its natural frequency of oscillation. The frequency changes proportionally with the deposited mass because the particles are collected on that filter progressively (transducer mass). It was work-operated with a sampling flow of 3 L min^{-1} and an operating temperature ranging between 25ºC and 50ºC.

Scanning mobility particle sizer (SMPS). This instrument model 3080 (TSI), measures size distributions in the 11 – 982 nm diameter range in real time with a 5 min scan rate. This system consists of a differential mobility analyzer (DMA), model 3081, and a condensation particle counter (CPC), model 3075A. The SMPS instrument is based on the principal of the mobility of a charged particle in an electric field. Sheath and aerosol sampling flows were 3 L min^{-1} and 0.30 L min^{-1}, respectively. Aerosol volume concentrations were converted to aerosol mass concentrations by assuming a density of 1 g cm^{-3}, spherical shapes and multi-charge correction for the condensed organic material. The correction of aerosol wall losses was performed by fitting the decrease of the integrated particle number concentrations over time to an exponential decay function after closure of the chamber to sunlight for each experiment.

Filter sampling system. It was also integrated in the smog chamber for off-line aerosol monitoring. This system consists of a filter holder (Millipore, Billerica, MA, USA) with a pre-heated 47 mm-diameter glass fiber filter. The sampling was performed at a flow of 60 L min^{-1}, when maximum aerosol formation was reached.

Moreover, methodologies for the determination of polycyclic aromatic compounds (PAHs), pesticides and hydroxycarbonyl compounds in particulate matter have been developed using a gas chromatograph – mass spectrometry (GC-MS). A TRACE-DSQ II instrument system was used (Thermo Fisher Scientific Co., Waltham, MA, USA).

2.3 Protocol experiments

Set-up experiments. Since reactor walls could be a source of gas and/or particles, due to the off-gassing of compounds, preliminary test are required before each run. For that, a chamber blank experiment has to be performed to assure the absence of possible artifacts. First, the high-volume chamber must to be flushed almost for 24 h with purified air. Secondly, the chamber must to be exposed to darkness to characterize the artifacts and to the sunlight to characterize the background levels.

Walls effect characterization. It is fundamental to understand the impact of reactor walls on gas-phase reactivity and particulate matter formation. The dilution is determined using SF_6 [9] as an non-reactive gas tracer. To address particles deposition wall loss, it is assumed a first order kinetics and the loss rate constant is dependent of particle size. Then, it is necessary to calculate loss coefficient for every measured size bin of the particle distribution as function of time [10].

Photolysis. The photolysis experiments are performed over approximately three hours in the middle of the day. To study the pure photolysis and to prevent any reaction with OH radicals, an excess of OH scavenger is added to the system.

Ozonolysis. The rate coefficient for the reaction of O_3 with precursor is determined by monitoring the loss of O_3 in the presence of a measured excess concentration of precursor, left in darkness for around 3 hours.

Photo-oxidations. The rate coefficient for the reaction of OH radicals with precursor is determined using a conventional relative rate method in which the concentrations of precursor and the reference compound are followed as function of time. There are several methods to induce OH generation: H_2O_2 photolysis - in the absence of NO_x -, the classic initial addition of NO and HONO photolysis - in the presence of NO_x -. The slowest, although roughly constant, OH generation is derived from the photolysis of H_2O_2 (6.04×10-6 s^{-1}). For the photo-oxidation experiments in the presence of NO_x, the radical source in the initial-NO experiments came from a heterogeneous reaction of NO_2 on the chamber wall [11], promoting a slower OH generation and a lower SOA production. Finally, in the photo-oxidation in the presence of NO_x, nitrous acid is added to introduce OH and NO_x by its photolysis. The rapid photolysis of HONO (1.44×10-3 s^{-1}) promoted a burst generation of OH which was continued by the radicals produced through recycling via NO_x/HO_x chemistry.

Study of degradation products. The experiments carried out for the physical and chemical characterization products are: photolysis, photo-oxidation in the presence or in the absence of nitrogen oxides and ozonolysis.

2.4 Experimental procedure

All degradation experiments were carried out by duplicate with a high reproducibility in the precursor consumption and degradation products formation. The experiments were under dry conditions - < 2% RH - or wet conditions adding pure water via sprayer - up to 55% - . The experiments were carried out in the absence. Precursors were introduced in the photo-reactor via heated air stream. The vapor pass through a FEP-Teflon transfer line and then injected into the chamber. Regarding the induced routes, H_2O_2 is introduced in the chamber with a sprayer, NO or NO_2 is added at 20 ppbV min^{-1} from an NO or NO_2 calibration bottle (99.999 % of 5000 ppm) and HONO was generated by a liquid-phase reaction between a 0.5 % $NaNO_2$ solution and a 30 % H_2SO_4 solution and transferred directly into the chamber via a stream of purified air. Ozone is added via a stream of gas produced in an ozonizator that converts O_2 in O_3.

Firstly, background data from all the instrumentation implied in the experiments is taken. The immediately analysis of this data informed us the state of the chamber initial conditions to continue or in the opposite case to abort the experiment. Later, all the reactants were introduced into the chamber and mixed with high power fans for 10 min before exposing them to sunlight. After that, all the instrumentation takes a sample to have the information of the mixture studied just before to start the degradation process. When all the samples were taken, the sunlight exposure was conducted or ozone was generated until the formation of secondary products in the gas phase or aerosol mass concentration reached the steady-state. Subsequently, the home was closed and the measurements continued for at least one hour to observe the aerosol decay behavior. Finally, particles were collected at maximum aerosol formation.

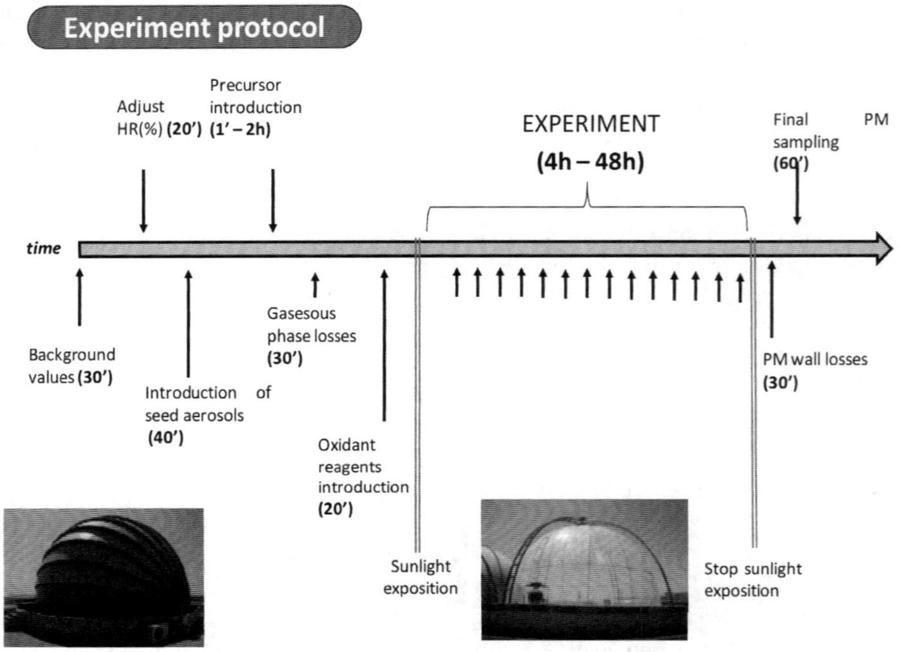

Figure 2. Time table of general experiment protocol.

3. Results

3.1 Aromatic hydrocarbons

Benzene is a major aromatic hydrocarbon air pollutant with an emission rate of 11 Tg year-1, and it plays a critical role in atmospheric chemistry. Atmospheric studies on benzene acquired greater relevance when this volatile organic compound (VOC) became established as a petrol additive, thus increasing its direct emission. Also, it is the only one aromatic hydrocarbon which ambient level is legislated - European Directive 2000/ 69/EC - . An example of experiment with benzene is herein showed. The experiments consisted of the photo-decomposition of benzene (1500 ppbV) under dry conditions in the absence of inorganic seed aerosol. The main gaseous products were phenol, ozone, catechol, formic acid, nitric acid and formaldehyde. Other minority gases were glyoxal, maleic anhydride, 4-nitrophenol and 2-nitrophenol (0.1-20 ppbV). Also, a significant amount of aerosol was obtained. In NOx-absence experiments, the beginning of aerosol nucleation required a sunlight exposure time around 2h, while for initial-NO conditions, the delay was around 1h, being the existence of a delay (or induction period) between the start of oxidation and SOA formation has also been observed for other aromatic hydrocarbons. Under the addition of HONO, aerosols were formed immediately after the exposition of the pollutant to sunlight radiation. Aerosol yields between 2 – 14%

IOP Conf. Series: Earth and Environmental Science **28** (2015) 012004 doi:10.1088/1755-1315/28/1/012004

were obtained and all results indicated that the SOA formation process is NOx-dependent (see Figure 3).

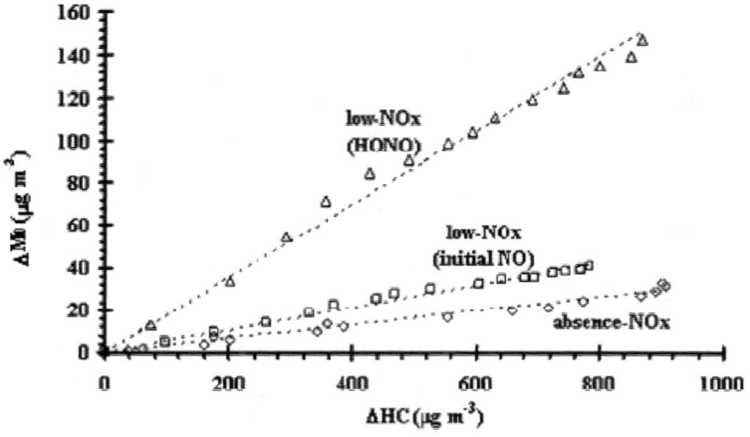

Figure 3. Particulate matter formation as function of precursor consumption under different NOx conditions.

3.2 Biogenic compounds

The emission of terpenes from vegetation is a major source of reactive organic species in the atmosphere with a rate in the order of 1014 g yr-1 on a global scale. Monoterpenes are important biogenic pollutants and play an important role in night-time atmospheric chemistry. Measurements performed in the last decade indicate that biogenic VOCs give the largest contribution to secondary organic aerosol formation. α-pinene, β-pinene and limonene are the most abundant terpenes [12]. In particular, the global production of SOA from the oxidation of biogenic compounds was estimated to be higher than 10 Tg yr-1. For that, the use of smog chambers as EUPHORE has a significant relevance since so the decay of terpenes so the formation of degradation products can be followed with a high time resolution and high precision and accuracy. Figure 4 shows the typical profile of ozonolysis experiment comparing the three main emitted monoterpenes.

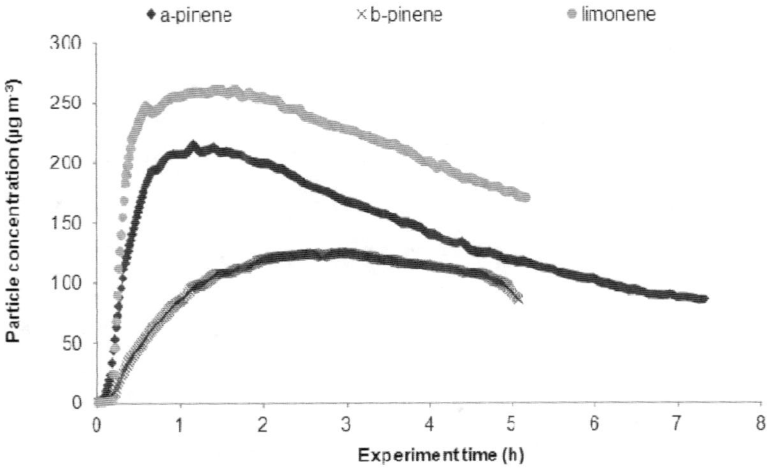

Figure 4. SMPS aerosol concentration profiles as a function of time.

The maximum mass concentrations (and aerosol yield) were 210 µg m-3 (21 %), 125 µg m-3 (14 %) and 250 µg m-3 (65 %), from α-pinene, β-pinene and limonene, respectively. These results agree with previously reported studies investigating SOA products from ozonolysis reactions of biogenic precursors in different-volume reaction chambers or SOA models [13]. The reproducibility of the aerosol formation obtained from replicates of ozonolysis experiments was also confirmed. The inter-day relative standard deviations of aerosol measurements were 5.0 %, 4.1 % and 1.8 % for α-pinene, β-pinene and limonene reactions, respectively. The main particulate matter products of a-pinene ozonolysis were pinic acid, pinonaldehyde, pinonic acid, norpinonic acid and hydroxypinonic acid. The main products of b-pinene were norpinone, pinonic acid, pinic acid, 2-hydroxynorpinone, norpinonic acid and norpinic acid. Finally, the main products of limonene were limononaldehyde, limononic acid, 7-OH-limonoaldehyde and keto-limononaldehyde,

3.2 Pesticides

Pesticides are extensively used in agriculture, gardening and a variety of other household applications. Its intensive use led to ubiquitous contamination of exposed and remoted areas, not only water or soil, but also in the atmosphere. However, studies of the gas-phase degradation of pesticides are problematic because of their low vapor pressures. One advantage of EUPHORE facilities is that compounds can be introduced into the chamber. Figure 5 shows an example of experimental data collected from degradation experiment of chlorpyrifos.

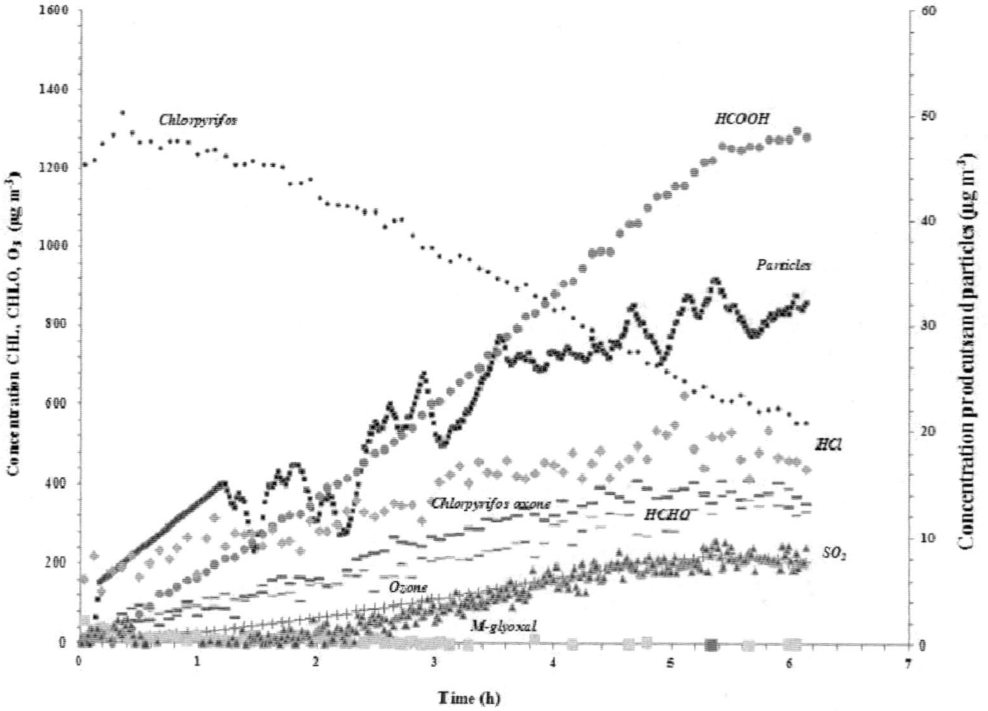

Figure 5. Experimental data collected for the photo-oxidation of chlorpyrifos.

The main degradation products were 3,5,6-trichloropyridine-2,4-diol, ethyl dihydrogen phosphate, o-ethyl o-(3,5,6-trichloro-4-hydroxypyridin-2-yl) hydrogen phosphoro-thioate, ethyl 2-oxoethyl hydrogen phosphate, 3,5,6-trichloropyridin-2-yl dihydrogen phosphate and, o -(3,5,6-trichloropyridin-2-yl) dihydrogen phosphorothioate.

4. Discussion and conclusion

The EUPHORE facility was designed to bring new insights on atmospheric photochemistry by providing new, precise and comprehensive data relevant to gas phase chemistry as well as to aerosol chemistry and physics at lower simulated pollutant concentrations. The data set from this chamber has been widely applied.

Furthermore, secondary aerosol yield studies from a known chemical system has shown good reproducibility and good agreement with previous studies indicating that this chamber is also well suited for studies of secondary aerosol formation and demonstrates our ability to accurately and precisely measure SOA formation potentials.

EUPHORE chamber is now integrating experimental simulations that will permit a significantly better integration of multiphase processes in more atmospherically representative conditions.

Acknowledgements
The research leading to these results received funding from the Spanish Ministry of Economy and Competitivity (project IMPLACAVELES: CGL2013-49093-C2-1-R), and Generalitat Valenciana for the DESETRES-Prometeo II project.

References
[1] Finlayson-Pitts B J, Pitts JN. Chemistry of upper and lower atmosphere. Academic Press, 2000.
[2] Borrás E, Tortajada-Genaro L A. 2012a *Atmos. Environ.* **47** 154
[3] Christoffersen T S, Hjorth J, Horie O, Jensen N R, Kotzias D. Molander LL, Neeb P, Ruppert L, Winterhalter R, Virkkula A, Wirtz K, Larsen B R 1998 *Atmos. Environ.* **32** 1657
[4] Tortajada-Genaro L A, Borrás E 2011 *J. Environ. Monit.* **13** 1017
[5] Wang J, Doussin J F, Perrier S, Perraudin E, Katrib Y, Pangui E, Picquet-Varrault B 2011 *Atmos. Meas. Tech. Discus* **4** 315
[6] Carter W P L, Cocker III D R , Fitz, D R Malkina I L, Bumiller K, Sauer C G, Pisano J T, Bufalino C, Song C 2005 *Atmos. Environ* **39** 7768
[7] Vivanco M G, Santiago M, Martínez-Tarifa A, Borrás E, Ródenas M, García-Diego C, Sánchez M 2011. *Atmos. Environ* **45** 708
[8] Vera T, Mñoz A, Rodenas M, Vázquez M, Borras E, Marques M, Mellouki A, Treacy J, Sidebottom H 2011 Atmospheric fate of hymexazol (5-methylisoxazol-3-ol): Simulation chamber studies. *Atmos. Environ* **45** 3704
[9] Volkamer R, Klotz B, Barnes I. Imamura T, Wirtz K, Washida N, Becker K H, Platt U 2002 Phys. Chem. Chem. Phys **4** 1598
[10] Baltensperger U, Kalberer M, Dommen J, Paulsen D, Alfarra M R, Coe H, Fisseha R, Gascho A, Gysel M, Nyeki S, Sax M, Steinbacher M, Prevot A S H, Sjogren S, Weingartner E, Zenobi R 2005 *Faraday Discus* **130** 278
[11] Ng N L, Kroll J H, Chan A W H, Chhabra P S, Flagan R C, Seinfeld J H 2007 *Atmos. Chem. Phys* **7** 3922
[12] Kroll J, Chan A W H, L, Ng N L, Flagan R C, Seinfeld J 2007 *Environ. Sci. Tech.* **41** 3545
[13] Atkinson R, Arey J 2003 *Chemical Review* **103** 4605

IOP Conf. Series: Earth and Environmental Science **28** (2015) 012005 doi:10.1088/1755-1315/28/1/012005

Analysis of derived optical parameters of atmospheric particles during a biomass burning event. Comparison with fossil fuel burning

A Costa[1], S Mogo[1,2,3], V Cachorro[2], A de Frutos[2], M Medeiros[1], R Martins[1], J F López[2], A Marcos[2], N Marcos[4], S Bizarro[4] and F Mano[4]

[1] Departamento de Física, Universidade da Beira Interior, Covilhã, Portugal
[2] Grupo de Óptica Atmosférica, Universidad de Valladolid, Valladolid, España
[3] Instituto Dom Luiz, Lisboa, Portugal
[4] Baldios da Freguesia de Cortes do Meio, Covilhã, Portugal

E-mail: smogo@ubi.pt

Abstract. During the day November 26, 2014, a scheduled cleanup of the woods took place around the GOA-UVa aerosol measurement station located at the campus of the University of Beira Interior (40° 16' 30" N, 7° 30' 35" W, 704 m a.s.l.), Covilhã, Portugal. This cleanup included excessive vegetation removal during the morning, using fossil fuel-burning machinery, and burning of the vegetation during the afternoon. *In situ* measurements of aerosol optical properties were made and this study aims the characterization of the evolution of aerosol properties during the day. The optical parameters were monitored using a 3-wavelength nephelometer and a 3-wavelength particle soot absorption photometer. Selective sampling/exclusion of the coarse particles was done each 5 minutes. The scattering and absorption Ångström exponents as well as the single scattering albedo were derived and fully analyzed. The scattering and absorption coefficients increased dramatically during the event, reaching values as high as 720.3 Mm^{-1} and 181.9 Mm^{-1}, respectively, for the green wavelength and PM10 size fraction. The spectral behavior of these parameters also changed wildly along the day and an inversion of the slope from positive to negative in the case of the single scattering albedo was observed.

1. Introduction

Biomass burning (BB) and fossil fuel burning (FFB) are strong sources of black carbon (BC) emissions into the atmosphere [1–4]. Despite their short lifetime, BC particles present strong effects. They are considered the strongest absorber of solar radiation in the atmosphere, interfering with the radiative balance of the planet and they are mainly fine fraction particles, inhalable by the human respiratory system and affecting human health. Studies that analyze the optical, chemical and microphysical properties of aerosols allow pave the way for understanding the contribution of these particles in radiative terms for the planet, human health and visibility [5, 6]. So, many authors direct their research to the characterization of aerosols that are released during the burning processes [1–3, 5–7]. However, due to the aggregation state of the particles (internal or external mixing), the study of all parameters associated with this type of particles becomes complex. Also, these particles, resulting from incomplete combustion processes, are

IOP Conf. Series: Earth and Environmental Science **28** (2015) 012005 doi:10.1088/1755-1315/28/1/012005

transported long distances in the atmosphere. Thus, the effects associated are not limited to the zone where burning occurred [7].

As Portugal has extensive forest areas, biomass burning is usually associated with forest fires, however, scheduled biomass burn activities are also a resource in agriculture for clearing of vegetation. Thus, such episodes can occur regularly. Given these facts and the importance of aerosol study in episodes with these features, this paper aims to present the evolution of the optical properties of aerosols during a scheduled cleanup of the woods that includes FFB as well as BB. The data is collected by the GOA-UVa *in situ* measurement station, located on the campus of the University of Beira Interior (UBI). This work is the result of the collaboration between the Group of Atmospheric Optics of the University of Valladolid (GOA-UVa) and UBI.

2. Methods

2.1. Sampling location

The events analyzed in this work were registered in Covilhã, a town located in central Portugal with $\sim$ 36700 inhabitants. Located in the proximity of the mountains (Serra da Estrela, 2000 m a.s.l.), the town towers between 450 and 800 m a.s.l. In the immediate vicinity, the mountains are enveloped by forested areas dominated by varieties of pine and chestnut trees, and the valley is dominated by agriculture in small farms. *In situ* measurements of aerosol properties, are continuously being collected since July 2013 in the GOA-UVa aerosol measurement station located at the campus of UBI, outside the town but 1 km straight line from the city center.

2.2. Instrumentation and data processing

For transporting aerosols from the atmosphere to the instruments, an inlet protected with a rain cap is used. The cut-off diameter of the inlet nozzle and sample transport line is approximately 10 μm and the airflow through the sampling line is divided into several separate flows and directed toward individual instruments. The collecting system includes a valve for selective sampling/exclusion of the coarse particles, which was done each 5 minutes in such a way that during 5 minutes only the fine fraction (PM1) reaches the instruments and the following 5 minutes the complete PM10 fraction reaches the instruments.

The scattering coefficients, σ_s, were measured at three wavelengths (450, 550 and 700 nm) using an integrating nephelometer (model 3563, TSI). The instrument is described in detail by [8] and [9]. The averaging time was set to 5 min. The zero signal was measured once per hour. The data was corrected for truncation errors according to [9].

The light absorption coefficients, σ_a, were measured at three wavelengths (470, 522 and 660 nm) using a particle soot absorption photometer (PSAP, Radiance Research). The instrument is described in detail by [10] and [11]. The averaging time used for the PSAP was 60 s, and the filter was replaced whenever the amount of transmitted light reached 70 % of the initial intensity. The data was corrected for the loading of particles on the filter, the amount of light scattered by the particles, the flow rate and the spot size according to [10] and [11].

The corrected aerosol scattering coefficients were extrapolated to the working wavelengths of the PSAP using the Ångström exponents.

The aerosol single-scattering albedo (SSA), ω_0, is obtained as the ratio between the aerosol light scattering and the extinction:

$$\omega_0 = \frac{\sigma_s}{\sigma_s + \sigma_a}. \tag{1}$$

To investigate the wavelength dependence of the optical parameters, we calculated the scattering/absorption/SSA Ångström exponent. This parameter is commonly used for the analysis of the variations of the spectral shape of the optical parameters and is defined as the negative slope of the logarithm of the optical parameter as a function of the wavelength. It is given by:

$$\sigma_x = K\lambda^{-\alpha_x}, \tag{2}$$

where x may represent the scattering coefficient, the absorption coefficient or the single scattering albedo. Using only one pair of wavelengths, the Ångström exponent can be obtained from the following simplified formula:

$$\alpha_x(\lambda_1, \lambda_2) = -\frac{\log(\sigma_x(\lambda_2)/\sigma_x(\lambda_1))}{\log(\lambda_2/\lambda_1)}. \tag{3}$$

Data is available for the day November 26, 2014 and the statistical data for all instruments were calculated based on 5-min averages.

3. Results and discussion

Figure 1(a) shows the evolution of the absorption and scattering coefficients, as well as single scattering albedo throughout the day. Figure 1(b) shows the evolution of the Ångström exponents. The analysis of the data series was divided into three stages: during the biomass burning episode, fossil fuel burning episode and rest of day. The statistical measures such as mean, standard deviation, maximum and minimum values are presented in table 1.

Table 1. Evaluation of the overall ranges and arithmetic mean values of the absorption/scattering coefficients and SSA as well as the respective Ångström exponents in PM1 and PM10 fractions. FFB: fossil fuel burning. BB: biomass burning. B, G ,R: blue, green and red wavelengths. Statistical values are presented for all the day, excluding the time of the events. For the BB and FFB events, only the peak value is presented.

PM10									PM1					
	rest of the day			FFB	BB				BB	FFB	rest of the day			
StD	max	min	mean	peak	peak				peak	peak	mean	min	max	*StD*
11.3	56.2	8.1	22.9	66.6	879.4	B		B	712.3	74.1	21.6	8.3	45.8	10.4
9.4	48.1	6.2	18.3	52.6	720.3	G	σ_s (Mm^{-1})	G	605.5	58.8	17.0	6.2	35.8	8.4
7.4	38.2	4.0	12.9	35.6	516.3	R		R	463.6	40.8	11.7	3.9	29.1	6.1
6.7	31.3	2.6	10.1	58.2	212.9	B		B	681.0	63.2	10.1	3.4	40.9	7.0
5.4	22.6	2.4	8.2	48.8	181.9	G	σ_a (Mm^{-1})	G	490.4	52.4	8.0	2.7	31.7	5.4
4.1	17.8	2.1	6.6	39.1	145.1	R		R	309.3	41.9	6.0	2.0	22.6	4.0
0.1	0.8	0.5	0.7	0.3	0.1	B		B	0.1	0.3	0.7	0.4	0.9	0.1
0.1	0.8	0.4	0.7	0.3	0.1	G	ω_0	G	0.2	0.3	0.7	0.5	0.9	0.1
0.1	0.8	0.4	0.7	0.3	0.2	R		R	0.2	0.2	0.7	0.5	0.9	0.1
0.1	2.2	1.6	1.9	1.9	2.5	B-G		B-G	2.5	2.0	1.9	0.8	2.3	0.3
0.3	2.4	0.5	1.8	1.9	2.5	B-R	α_s	B-R	2.5	2.0	1.9	0.7	2.5	0.3
0.2	2.5	1.4	1.9	2.0	2.6	G-R		G-R	2.5	2.0	1.9	0.7	2.5	0.3
0.3	2.8	1.2	1.8	1.7	2.7	B-G		B-G	3-0	1.7	1.8	1.2	3.2	0.4
0.2	2.5	1.1	1.6	1.5	2.5	B-R	α_a	B-R	2.8	1.5	1.6	1.2	2.7	0.3
0.2	2.3	1.1	1.4	1.4	2.5	G-R		G-R	2.7	1.3	1.4	1.1	2.4	0.2
0.1	0.3	0.1	0.04	0.03	0.9	B-G		B-G	0.9	0.1	0.03	0.5	0.3	0.1
0.1	0.4	0.3	0.1	0.1	0.8	B-R	α_{ω_0}	B-R	0.7	0.2	0.1	0.4	0.4	0.1
0.1	0.4	0.03	0.2	0.1	0.7	G-R		G-R	0.6	0.2	0.2	0.3	0.4	0.1

The period of the BB episode is easily identified by the pronounced peak in figure 1(a). The FFB episode can also be detected mainly by increase of the absorption values in the morning. We will present values for one wavelength (green) for easiness on reading. The σ_s values registered

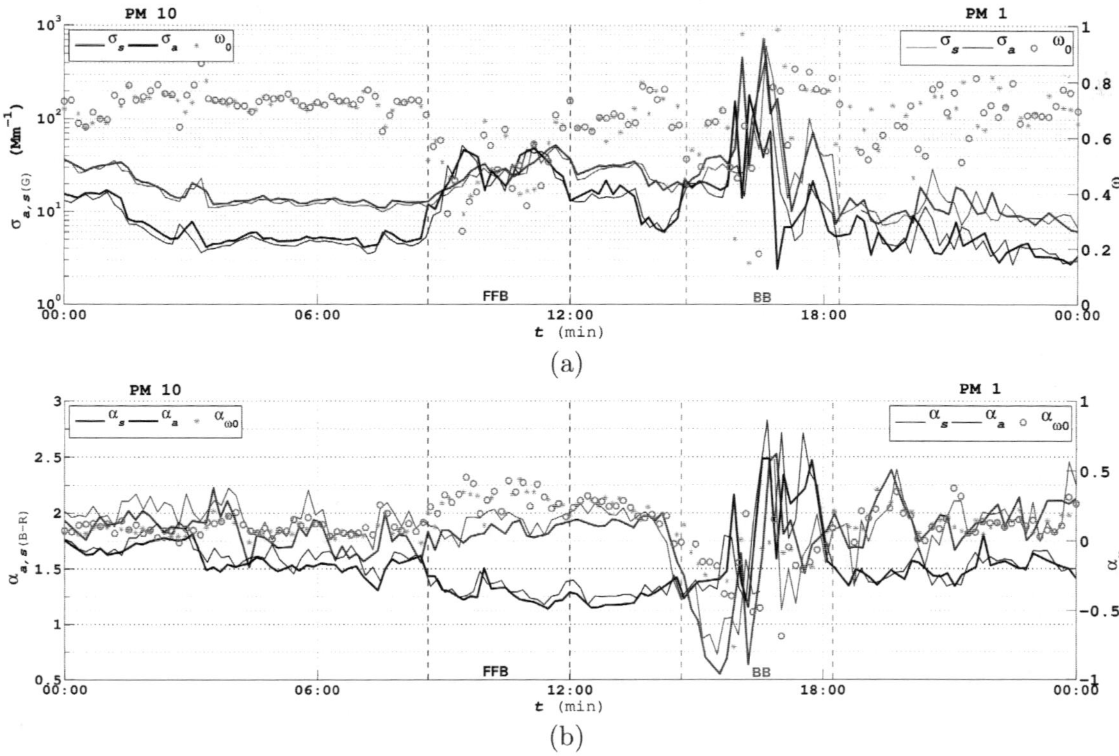

Figure 1. Temporal evolution of optical parameters during the day. FFB: fossil fuel burning. BB: biomass burning.

during the day without including the BB and the FFB events ranged from 6.2 to 48.1 Mm^{-1} (average 18.3 and standard deviation of 9.4 Mm^{-1}) on PM10. For the fine fraction, PM1, the σ_s ranged from 6.2 to 35.8 Mm^{-1} (average 17.0 and standard deviation of 8.4 Mm^{-1}). The σ_a values registered during the day without the events ranged from 2.4 to 22.6 Mm^{-1} (average 8.2 and standard deviation of 5.4 Mm^{-1}) on PM10. For PM1 the σ_s ranged from 2.7 to 31.7 Mm^{-1} (average 8.0 and standard deviation of 5.4 Mm^{-1}). The average ω_0 values observed in the period out of events was 0.7 (standard deviation of 0.1) for both fractions, PM10 and PM1. During the BB event the σ_s reached values of 720.3 and 605.5 Mm^{-1}, the σ_a reached values of 181.9 and 490.4 Mm^{-1}, the ω_0 decreased to values as low as 0.1 and 0.2, respectively for PM10 and PM1 fractions. During the FFB event the σ_s reached values of 52.6 and 58.8 Mm^{-1}, the σ_a reached values of 48.8 and 52.4 Mm^{-1}, respectively for PM10 and PM1 rations. The ω_0 decreased to values as low as 0.3.

During the BB event, for the pair of wavelengths B-R, α_s assumed extreme values (0.5 to 2.5) showing the presence of small and large particles throughout the burning event, figure 2(a). The α_a values were spread in the range 1.2 to 2.8 showing the presence of different chemical species during the event, figure 2(b). α_{ω_0} assumed negative values, which means an increase of the SSA along the spectrum, figure 2(c). Figure 2(d) shows a large range of values, however, the most extreme values can also be considered a characteristic of the BB.

During the FFB event, for the same pair of wavelengths B-R, α_s assumed values in the range 1.7 to 1.9 showing only slight variation in the size of particles during the event, figure 2(a). The α_a showed values very concentrated in the range 1.1 to 1.5 as a characteristic of the burning particles, figure 2(b). α_{ω_0} remained always positive showing a decrease of SSA with

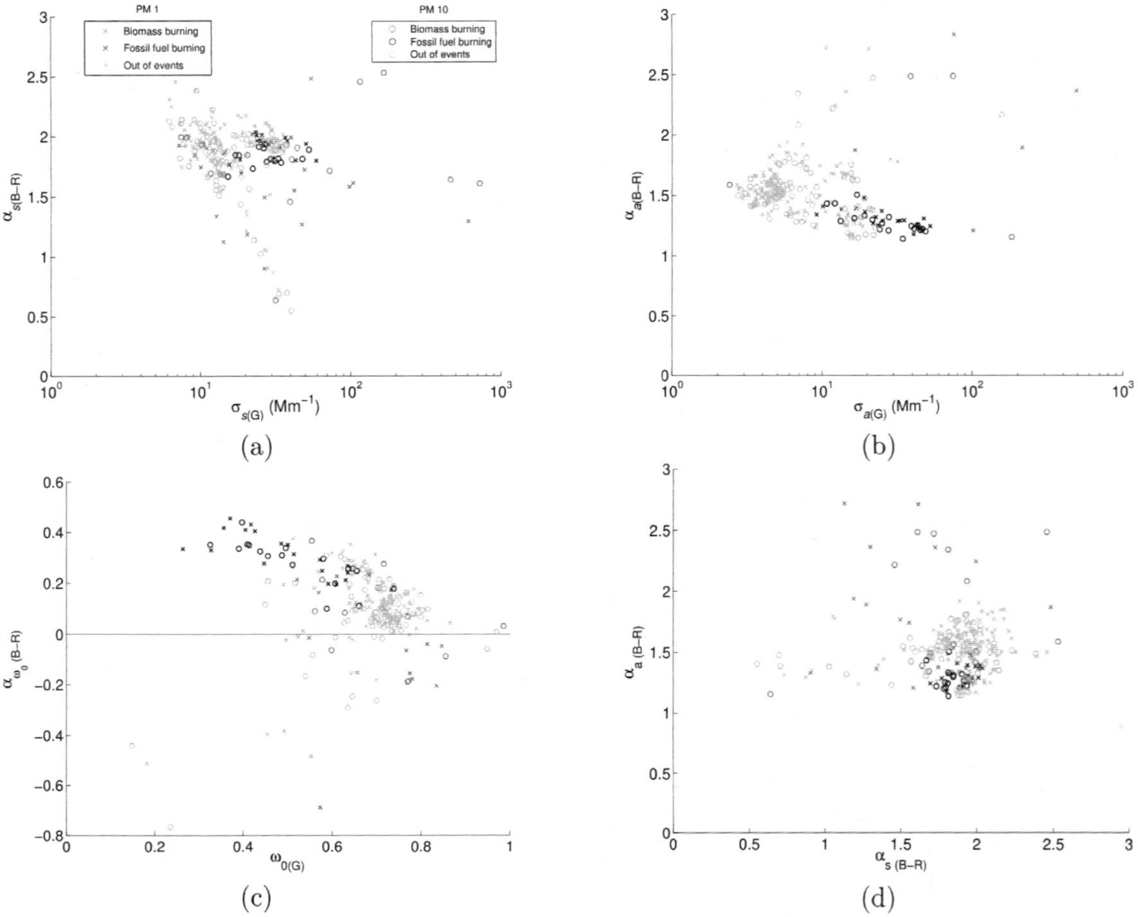

Figure 2. (a), (b), (c) Optical parameters as a function of the respective Ångström exponents. (d) Absorption versus scattering Ångström exponents.

the wavelength, figure 2(c). Figure 2(d) shows a very concentrated range of values that can be considered characteristic of FFB events.

The wavelength dependence of absorption is another way to understand the relative importance of biomass and fossil fuel contribution to BC. Figure 3 compares the normalized light absorption as a function of wavelength for the two events, showing higher spectral dependence during BB and softer dependence during FFB. This result agrees with the observations of other authors studying similar events [3].

4. Final comments

We studied the optical evolution of aerosols resulting from a biomass burning and a fossil fuel combustion events that occurred the same day. The study was conducted in central Portugal, in a region dominated by forest and agriculture in small farms. The primary optical measurements made were light absorption by particle soot absorption photometry and light scattering by nephelometry. Very high values were observed for the scattering and the absorption coefficients, which lead to low SSA values. The scattering and absorption Ångtröm exponents showed similar behavior, with higher values for the pair blue-green and smaller values for the pair green-red. During the biomass burning event, the Ångtröm exponent of the SSA assumed negative values, showing an increase of SSA along the spectrum. The fundamental insight gained from this

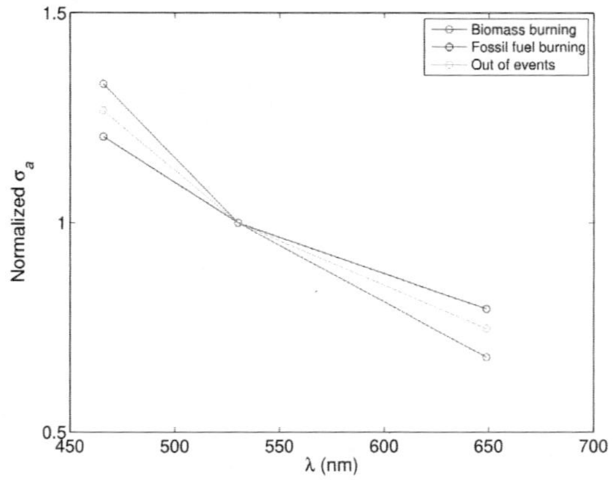

Figure 3. Comparison of normalized light absorption in aerosol samples collected during the biomass burning event, the fossil fuel burning event and non-event hours.

field study is the distribution of the scattering and absorption Ångtröm exponents during both events. While the biomass burning event is characterized by the most extreme values registered, the fossil fuel combustion event is characterized by a very concentrated range of values. Also, during biomass burning events, the spectral shape of the absorption coefficient drops quicker than in average conditions out of event. During fossil fuel combustion events, the spectral shape of the absorption coefficient drops more slowly than in average conditions out of event.

Given the fact that a large number of scheduled biomass burn activities are done as a resource in agriculture for clearing vegetation, in Portugal and other countries, it is desirable to evolve methodologies that systematically study these events, both quantitatively and qualitatively. Further the findings from our work establish the need to focus on assessing the effects of black carbon emitted from biomass burning and fossil fuel combustion to the radiative forcing, the human health, etc, in order to evaluate its potential impacts.

References

[1] Reid J S, Koppmann R, Eck T and Eleuterio D 2005 *Atmos. Chem. Phys.* **5** 799–825

[2] Adler G, Flores J M, Riziq A A, Borrmann S and Rudich Y 2011 *Atmos. Chem. Phys.* **11** 1491–1503

[3] Rehman I, Ahmed T, Praveen P, Karl A and Ramanathan V 2011 *Atmos. Chem. Phys.* **11** 7289–7299

[4] IPCC 2013 IPCC Fifth Assessment Report: Climate Change 2013 (AR5) Tech. rep. URL `\verbatim{http://www.ipcc.ch/report/ar5/wg1/#.UvkN-lhdLQw}`

[5] Leino K, Riuttanen L, Nieminen T, Maso M D, Vaananen R, Pohja T, Keronen P, Jarvi L, Aalto P, Virkkula A, Kerminen V M, Petaja T and Kulmala M 2014 *Boreal Env. Res.* **19** 275–292

[6] Tiwari S, Pipal A, Srivastava A, Bisht D and Pandithurai G 2015 *Environ. Sci. Pollut. Res.* **22** 2846–2855

[7] Sciare J, Oikonomoul K, Favez O, Liakakou E, Markaki Z, Cachier H and Mihalopoulos N 2008 *Atmos. Chem. Phys.* **8** 5551–5563

[8] Anderson T, Covert D, Marshall S, Laucks M, Charlson R, Waggoner A, Ogren J, Caldow R, Holm R, Quant F, Sem G, Wiedensohler A, Ahlquist N and Bates T 1996 *J. Atmos. Ocean. Technol.* **13** 967–986

[9] Anderson T and Ogren J 1998 *Aerosol Sci. Technol.* **29** 57–69

[10] Bond T, Anderson T and Campbell D 1999 *Aerosol Sci. Technol.* **30** 582–600

[11] Virkkula A, Ahlquist N, Covert D, Arnott W, Sheridan P, Quinn P and Coffman D 2005 *Aerosol Sci. Technol.* **39** 68–83

IOP Conf. Series: Earth and Environmental Science **28** (2015) 012006 doi:10.1088/1755-1315/28/1/012006

Analysis of atmospheric vertical profiles in the presence of desert dust aerosols

M J Costa[1], M A Obregón[2], S Pereira[2], V Salgueiro[2], M Potes[2], F T Couto[2], R Salgado[1], D Bortoli[2], and A M Silva[2]

[1]Instituto de Ciências da Terra - Polo de Évora, Departamento de Física, Escola de Ciências e Tecnologia, Universidade de Évora, Évora, Portugal
[2]Instituto de Ciências da Terra - Polo de Évora, Departamento de Física, Instituto de Investigação e Formação Avançada, Universidade de Évora, Évora, Portugal

E-mail: mjcosta@uevora.pt

Abstract. The present work aims at studying a very recent episode of desert dust transport that affected Iberia in mid May 2015. The dust aerosols were detected over Évora, where a varied set of instrumentation for aerosol measurements is installed, including: a CIMEL sunphotometer integrated in AERONET, a Raman Lidar and a TEOM monitor, as well as ceilometer and a microwave radiometer (profiler). The aerosol occurrence, detected using the columnar, vertically-resolved and in situ measurements, was characterized by a fairly high aerosol optical thickness that reached a value of 1.0 at 440 nm and showed mass concentration peaks at the surface of the order of 100 $\mu g/m^3$. Subsequently, the tropospheric vertical profiles of humidity and temperature obtained with the passive microwave (MW) radiometer are analysed in order to distinguish possible modifications that can be connected with the transport of desert dust. Modelling results are also examined and the total, SW and LW radiative forcings are investigated, taking into account the different vertical profiles obtained during the desert dust occurrence. It is found that the differences in the atmospheric profiles mostly affect the LW radiative forcing, with an underestimation of about 30% when the actual vertical profile is not considered.

1. Introduction

Solar radiation is the main source of energy for the Earth-atmosphere system. It is well known that any change in the atmospheric composition would significantly affect the radiative budget and, as a result, the global temperature of the Earth. Important components of this system are atmospheric aerosols, interacting directly and indirectly with solar and terrestrial radiation. Desert dust aerosols from North Africa are frequently transported towards the Iberian Peninsula, driven by propitious synoptic conditions [1-3]. These episodes are most frequent during summertime [4-6], when air masses from Northern Africa are favored to reach higher latitudes. However there are still many uncertainties and accurate reliable measurements and analyses are demanded to reduce those uncertainties [7]. This sorts Iberia as a privileged location to study desert dust aerosols and the interaction of these particles with radiation and clouds. In order to quantify the effects of atmospheric aerosols on solar and terrestrial radiation, the concept of aerosol radiative forcing (ARF) is generally used. This variable indicates the magnitude of change in the radiative balance at a given level due to changes in aerosol physical/optical properties.

The aim of this work is to study an intense episode of desert dust transport that affected Évora site, Portugal, in mid May 2015. The vertical distribution of atmospheric variables as temperature and water vapor, obtained from a microwave radiometer profiler, is investigated in order to identify features that may be associated with the desert dust event. SW and LW radiative forcings are examined as well, taking into account the different vertical profiles obtained during the desert dust occurrence.

The work is organized as follows: after the introduction, a brief description of the site, instrumentation used and a characterization of the desert dust transport are presented in section 2; the method is described in section 3 followed by the results and discussion in section 4. Finally, conclusions are summarized in section 5.

2. Site, instrumentation and characterization of the desert dust transport

Simultaneous measurements of several atmospheric quantities are taken in Évora, at the Atmospheric Physics Observatory of the University of Évora (Institute of Earth Sciences - ICT) (38°34' N, 7°54' W, 293 m a.s.l). Mass concentration measurements at the surface are measured with a TEOM (Tapered Element Oscillating Microbalance) Series1400a monitor, installed in the site since January 2006, provided with a PM_{10} sampling head to remove particles with an aerodynamic diameter larger than 10 μm. The aerosol optical thickness (AOT) is measured with a CIMEL CE-318 sunphotometer, which is integrated in AERONET [8]. Figures 1 and 2 show the hourly mass concentration measurements and the AOT at 440 nm, respectively, between 10 and 16 May 2015. It can be noted that high values of mass concentration and AOT were measured on 12 and 13 May, which gradually decreased during the 14 May. These high values were accompanied by a strong decrease of the Angstrom exponent (AE) in the same days (also represented in the right hand axis of Figure 2) that gradually increased during the 14 May. The high values of mass concentration at the surface, together with high AOT and low AE values are consistent with the presence of desert dust in the region.

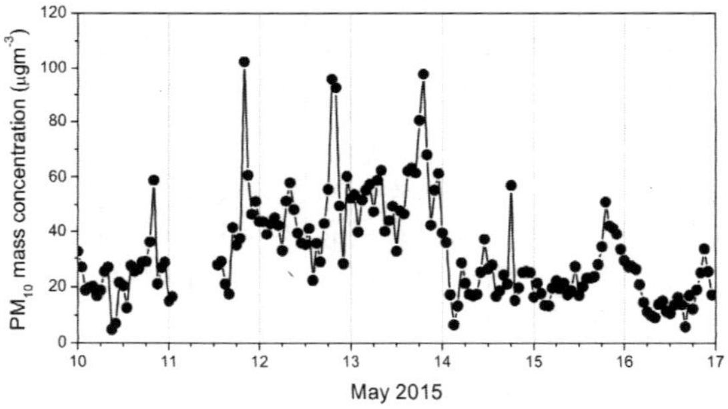

Figure 1. Hourly mass concentration measurements at the surface taken with a TEOM in Évora between 10 and 17 May 2015.

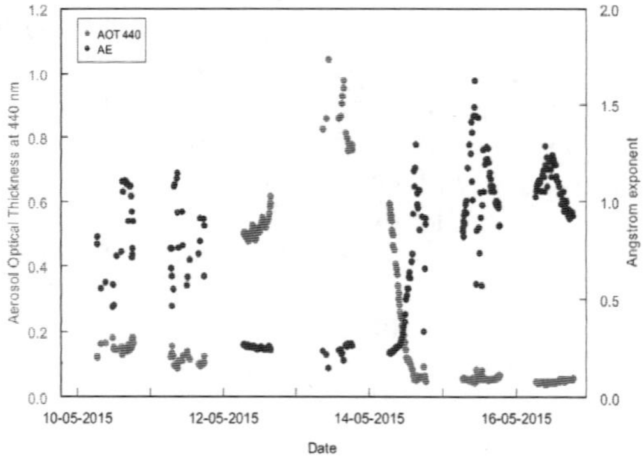

Figure 2. AERONET AOT at 440 nm (left hand axis) and Angstrom exponent between 440 and 870 nm (right hand axis), between 10 and 17 May 2015, over Évora.

The active remote sensing measurements, obtained with a Raman lidar system and a ceilometer (Vaisala CL31), are shown respectively in Figures 3 and 4 and illustrate the vertical distribution of aerosols (dust).

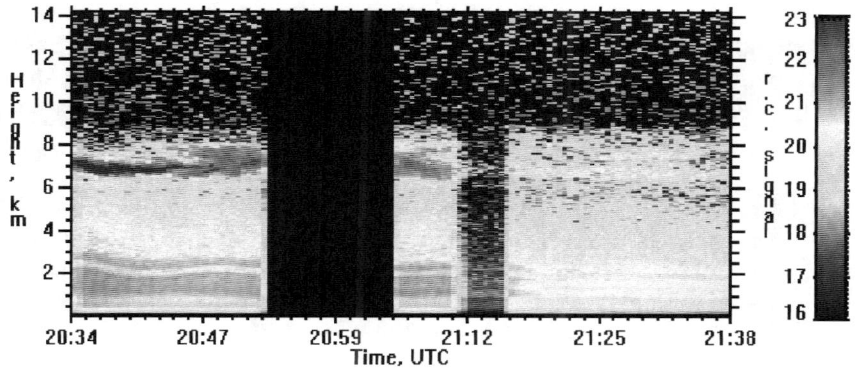

Figure 3. Range corrected signal at 532 nm, measured with the Évora Raman lidar system in the evening of 13 May 2015.

The lidar range corrected signal at 532 nm (Figure 3) shows a widespread aerosol layer extending from the lower levels up to about 4-5 km, with a lower signal for increasing height, which corresponds to lower aerosol concentrations. The stronger signal above the dust layer, between 7 and 8 km, corresponds to cirrus clouds. The orange vertical stripes depicted in Figure 4 indicate the time span with lidar measurements (Figure 3), although the weaker ceilometer laser does not allow having much information above 2 km, a diminution of the signal may be distinguished; on the other hand, the typical lidar overlap prevents this instruments to measure at very low altitudes (first few hundreds of meters), where the ceilometer is a good option. The red features in the first levels above the surface (Figure 4) are related with technical issues of the instrument. A good choice is the combination of both in order to gather as much vertical information as possible. The ceilometer backscattering profiles allow for distinguishing an aerosol layer on 12 May that separates from the boundary layer around noon and is visible during the next day and until the morning of 14 May, which probably corresponds to the dust aerosols.

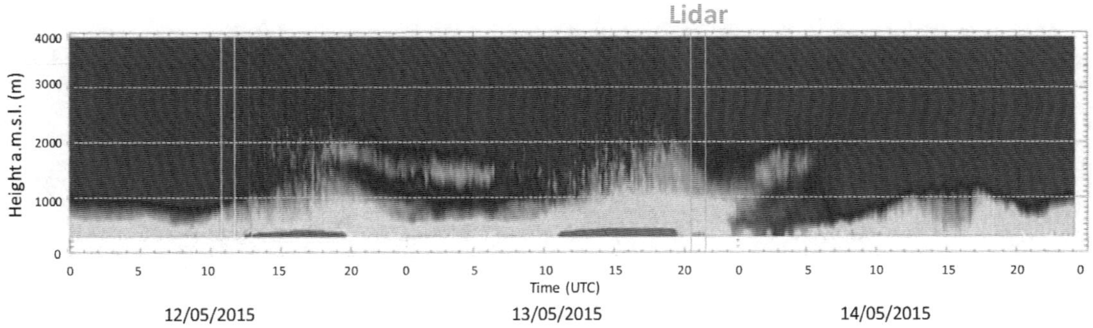

Figure 4. Ceilometer (CL31) backscattering profiles in Évora, between 12 and 14 May 2015. The orange vertical stripes indicate the time span with Lidar measurements (Figure 3).

The measurements presented before hinting at the desert dust intrusion, are corroborated by the back trajectory analysis presented in Figure 5, corresponding to the 120-h back trajectories arriving at Évora at 500, 2000 and 4000 m between 11 and 14 May 2015 obtained from HYSPLIT model (Hybrid Single-Particle Lagrangian Integrated Trajectory) version 4 [9-10]. The back trajectories show that on 11 May the circulation is from west at middle-levels (North Atlantic Ocean), but from 12 to 14 May the higher trajectories are originated in North Western African (NWA) regions, maybe transporting desert dust into Iberia. At the surface it seems there may have been also some intrusion (13 May), although it is not so evident.

In agreement with the back trajectory analysis, the transport of dust from NWA regions toward Iberian Peninsula was favored by the synoptic environment during the period (charts not shown here). The dust was transported by the southwesterly flow at middle levels established due the presence of a ridge extending from North Africa to Central Europe (11 May). The development of a relatively deep high pressure center was observed near the Strait of Gibraltar during 12 and 13 May, bringing NWA inland air into the southwest flow. At the surface, the development of a cyclone near the Azores was observed during 11 and 12 May. The cyclone developed along a semi-stationary frontal zone on 11 May associated to a low pressure system centered approximately at 55ºN and 20ºW. After the dissipation of the system on 13 May, the Azores anticyclone retook its normal position, well configured at the surface during 14 May, and the general westerly flow over Portugal was restored, thus ending the southwesterly transport of dust from NWA.

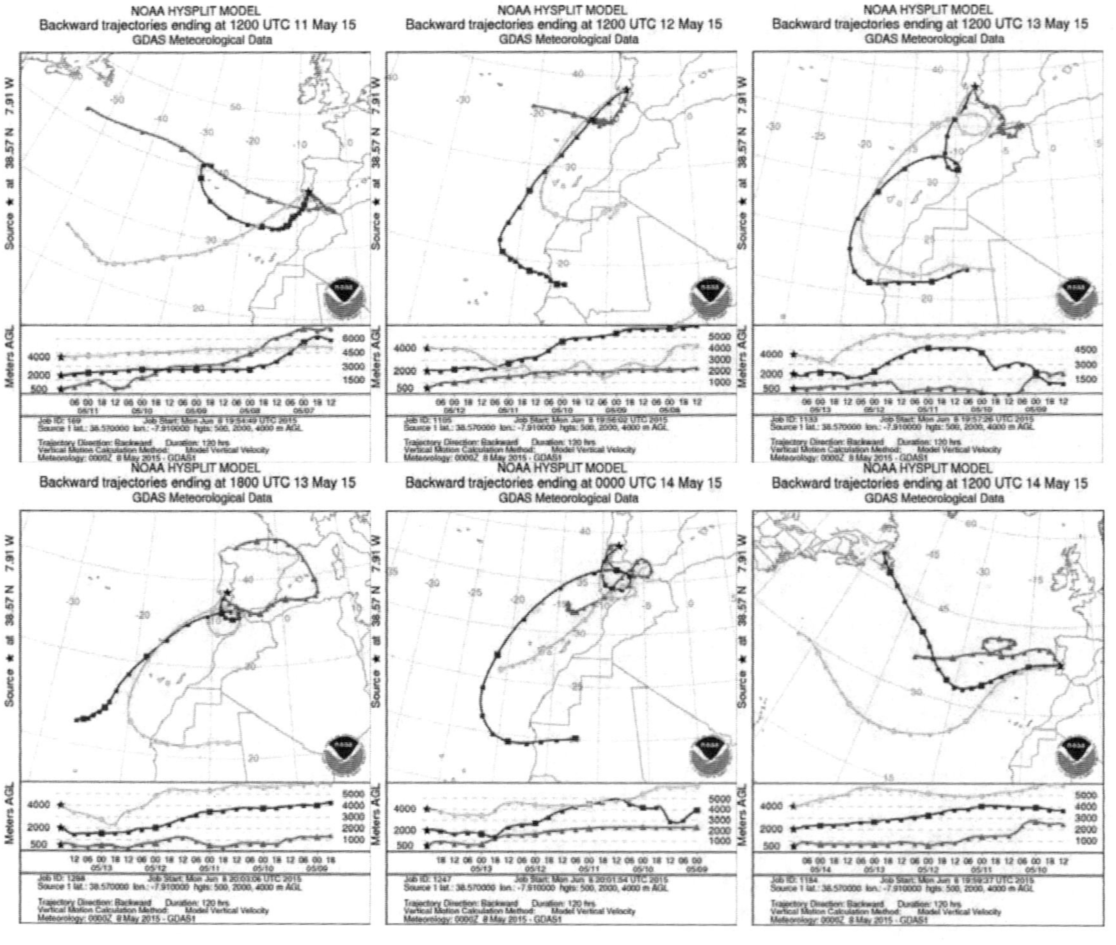

Figure 5. 120-h back trajectories arriving at Évora at 500, 2000 and 4000 m during the period 11-14 May 2015.

The atmospheric vertical profiles of temperature and humidity used in this work are obtained with the RPG-HATPRO humidity and temperature profiling passive microwave radiometer (MWR) [11], installed in Évora since September 2014. The MWR also measures liquid water path (LWP), integrated water vapor (IWV), wet and dry delay and stability indices. It has two bands, 22-31 GHz (7 channel filter bank humidity profiler and LWP radiometer) and 51-58 GHz (7 channel filter bank temperature profiler) and presents zenith and azimuth scanning capabilities. Attached to the MWR there are also a meteorological station and an infrared radiometer providing cloud base detection.

3. Method

The analysis of the atmospheric vertical profiles of temperature and humidity (absolute and relative) obtained with the MWR is done through the inspection of these profiles in the days before, during and after the desert dust transport occurrence. The profiles are intercompared and checked against a widely used climatological atmospheric profile [12].
As mentioned above, the aerosol effects on solar and terrestrial radiation are usually quantified through the so called aerosol radiative forcing (ARF), calculated in this work at the surface level. The ARF is defined here (similarly to AERONET) as the difference between the downward global solar irradiance at the surface in the presence of aerosols, I_d, and the same quantity in background/baseline conditions, I_d^0, (Eq. 1), assuming that, in background/baseline conditions, the aerosol optical thickness is null, as:

$$ARF = I_d - I_d^0 \qquad (1)$$

The hourly mean values of ARF, and subsequently daily values of ARF, were calculated as well. For this, the daily averaged forcing is calculated by integrating the hourly data during the whole day (24 h):

$$DARF_{libRadtran} = \frac{\sum ARF_{hourly}}{24} \qquad (2)$$

The shortwave and longwave irradiances are simulated with the libRadtran radiative transfer model [12]. Version 1.7 of the libRadtran is used in this study with inputs of aerosol optical properties, columnar ozone, precipitable water vapor column and surface albedo. Aerosol properties and precipitable water vapor were obtained from AERONET measurements. Total ozone column was provided by the Ozone Monitoring Instrument (OMI). The surface albedo data have been obtained from the Surface and Atmospheric Radiation Budget (SARB) working group, part of NASA Langley Research Center's Clouds and the Earth's Radiant Energy System (CERES) mission. Other variables taken into account in setting up the model are the following: extraterrestrial irradiance values (obtained from [13]) and the radiative transfer equation solver (the discrete ordinate method [14], DISORT2 calculated with 16 streams, was used). The vertical profiles considered are those obtained from the MWR measurements and a climatological profile is also considered for comparison purposes.

4. Results and discussion

Figure 6 shows the atmospheric profiles obtained from the MWR measurements at 12:00 UTC, between 11 and 14 May 2015. Clearly there is a distinction for all variables, between the profiles obtained on 12-13 May (desert dust event) and the rest of the days. The temperature profiles on 12 and 13 May show higher values in the lower troposphere (up to about 4 km), whereas in the higher levels there is little difference between all MWR temperature profiles. The midlatitude summer (mls) temperature profile presents considerable differences with respect to the MWR temperature profiles, overestimating the temperature values above 5 km with respect to all MWR profiles and underestimating it below 3 km, especially with respect to 12 and 13 May MWR profiles. The absolute humidity on 12 and 13 May present lower values for lower altitudes, with a minimum at about 1 km, then the values increase becoming coincident with the rest of the days at 2 km, and from this altitude until 6 km the absolute humidity is higher than for the rest of the days. Concerning the mls absolute humidity profile, the values are practically coincident with those from the MWR obtained for 11, 14 and 15 May, only slightly higher between 1 and 3 km of altitude. The relative humidity shows differences as well, with the MWR profiles for 12 and 13 May reaching a minimum (less than 10%) at 1 km and then steadily increasing until about 4 km and from this height on becoming roughly constant. This behavior differs from the other 3 days (11, 14 and 15 May), when the decrease at lower levels is not so abrupt, neither the increase at higher levels.

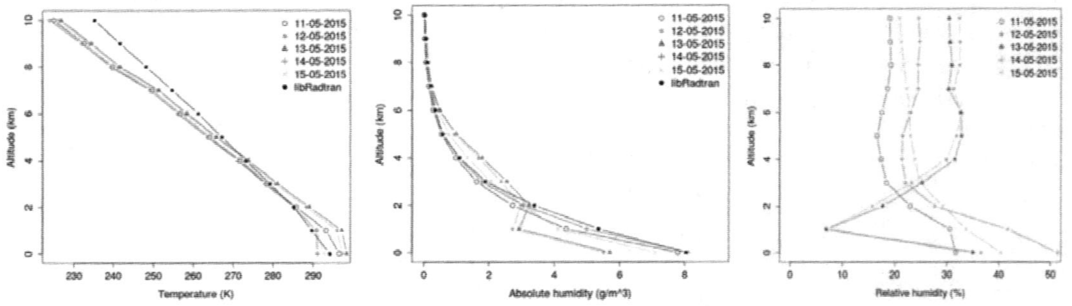

(a) (b) (c)

Figure 6. Atmospheric vertical profiles obtained from MWR measurements from 11 to 15 May 2015 at 12:00 UTC and mls profile contained in libRadtran database: a) temperature; b) absolute humidity; c) relative humidity.

The different behavior of temperature and humidity obtained for 12 and 13 May with respect to the other days is most probably related with the different circulation (coming from North Africa) that affected the south of Portugal during these days, when warmer air masses with a different vertical moisture distribution with respect to those reaching the site before and after, coming from west through the Atlantic (Figure 5), transported the desert dust particles. This low level warm air is rather dry below 2 km and above this altitude atmospheric moisture exceeds that of the other days analyzed. This could be connected with the differences in the low and mid-level trajectories, with the first more continental and the second over the Atlantic Ocean (see Figure 5).

Table 1. Daily mean SW and LW clear-sky downwelling irradiance at the surface for different vertical profiles of temperature and absolute humidity (mls: mls profile; MWR: profiles obtained from MWR measurements; MWR T: temperature profile obtained from MWR measurements and mls humidity profile; MWR H: mls temperature profile and humidity profile obtained from MWR measurements).

Vertical profile considered	SW downwelling irradiance at the surface (W/m^2)				LW downwelling irradiance at the surface (W/m^2)			
	11 May	12 May	13 May	14 May	11 May	12 May	13 May	14 May
mls	666.26	614.35	595.05	647.30	299.74	307.94	310.65	302.24
MWR	666.34	616.07	596.75	647.88	306.59	313.65	321.78	290.60
MWR T	666.16	614.36	595.09	647.16	306.78	318.72	326.70	293.56
MWR H	666.41	616.08	596.75	648.03	299.60	302.91	305.84	299.06

The daily mean SW and LW clear-sky downwelling irradiances at the surface were calculated from 11 to 14 May, considering different vertical profiles of temperature and absolute humidity. The results are summarized in Table 1, where simulations are identified as: mls for simulations using the mls profile; MWR: simulations using the profiles obtained from MWR measurements; MWR T: simulations considering the temperature profiles obtained from MWR measurements and the mls

IOP Conf. Series: Earth and Environmental Science **28** (2015) 012006 doi:10.1088/1755-1315/28/1/012006

humidity profile; MWR H: simulations using a mls temperature profile and the humidity profiles obtained from MWR measurements. In the case of SW radiation, there is very little change in the four types of simulation, although on 12 and 13 May a slightly higher change is observed. Accordingly, results suggest that, to some extent, the atmospheric moisture plays a more important role since MWR H simulations are nearer MWR values than MWR T. Nevertheless, results show that the atmospheric profile does not play a central role in the clear-sky SW irradiance at the surface. As for LW irradiance, the differences obtained in the four simulations for each day demonstrate the importance of using the actual atmospheric profile. In addition, it can be also concluded for the results in Table 1 that in this case, the temperature profile seems to have greater importance since the results of MWR T are nearest to MWR (considered here the actual irradiance values).

Table 2. Daily mean SW, LW and total aerosol radiative forcing from 11 to 14 May 2015, considering two types of atmospheric vertical profiles: climatological and microwave measurements.

	Aerosol radiative forcing (W/m^2)					
	SW		LW		Total	
Date	mls atmospheric vertical profiles	MWR atmospheric vertical profiles	mls atmospheric vertical profiles	MWR atmospheric vertical profiles	mls atmospheric vertical profiles	MWR atmospheric vertical profiles
2015-05-11	-6.04	-5.97	+0.14	+0.20	-5.9	-5.77
2015-05-12	-34.57	-33.60	+3.77	+5.31	-30.80	-28.29
2015-05-13	-45.30	-44.33	+4.64	+6.89	-40.66	-37.44
2015-05-14	-15.05	-14.71	+1.24	+1.65	-13.81	-13.06

The daily mean SW, LW and total (SW plus LW) ARF values calculated from 11 and 14 May 2015 are presented in Table 2. The SW ARF presents the lowest values on 12 and 13 May, when the AOT values were higher (Figure 2). As expected due to the results in Table 1, the use of mls or MWR profiles does not change significantly the ARF obtained (change of about 2% on the 13 May), with a slight tendency of mls case to underestimate the SW ARF with respect to the MWR case. Nonetheless, the same cannot be concluded for the LW ARF, where meaningful underestimations are found due to the use of mls profile, reaching roughly 33% on the 13 May. These results reflect on the total ARF, causing an underestimation when the actual vertical profile is not considered.

5. Conclusions

The work is focused on the transport of Saharan desert dust that occurred in mid May 20015 and affected the south of Portugal. The episode is characterized using *in situ*, as well as passive and active remote sensing measurements. The tropospheric vertical profiles of humidity and temperature obtained with the passive microwave radiometer show that during the dust transport the atmospheric temperature is systematically higher than in the other days and the humidity is lower in the low atmosphere, becoming higher than the reference days for the higher tropospheric levels. The differences found are probably related with the different circulation that affected the south of Portugal during these days, bringing warm air from North Western Africa, with a different vertical moisture distribution with respect to those reaching the site before and after. The analysis of SW and LW dowelling irradiance and ARF at the surface level show that the differences in the atmospheric profiles mostly affect the LW, with an underestimation of about 30% in the ARF when the actual vertical profile is not considered.

Acknowledgments
This work was partially supported by FCT (Fundação para a Ciência e a Tecnologia) through grants SFRH/BD/88669/2012, SFRH/BPD/86498/2012, SFRH/BPD/81132/2011, SFRH/BPD/97408/2013 and SFRH/BD/81952/2011 and project FCOMP-01-0124-FEDER-014024 (PTDC/AAC-CLI/114031/2009) and FCOMP-01-0124-FEDER-029212 (PTDC/GEO-MET/4222/2012). The authors acknowledge the funding provided by ICT, under contract with FCT (the Portuguese Science and Technology Foundation). The authors gratefully acknowledge the NOAA Air Resources Laboratory (ARL) for the provision of the HYSPLIT transport and dispersion model and/or READY website (http://www.ready.noaa.gov) used in this publication. Thanks are due to AERONET/PHOTONS and RIMA networks for the scientific and technical support. CIMEL calibration was performed at the AERONET-EUROPE GOA calibration center, supported by ACTRIS under agreement no. 262254 granted by European Union FP7/2007–2013. The authors also acknowledge Samuel Bárias for maintaining instrumentation used in this work.

References
[1] Rodriguez S, Querol X, Alastuey A, Kallos G and Kakaliagou O 2001 Saharan dust contributions to PM10 and TSP levels in Southern and Eastern Spain, *Atmos. Environ.* **35** 2433–47
[2] Querol X, Rodriguez S, Cuevas E, Viana MM and Alastuey A 2002 Intrusiones de masas de aire africano sobre la Península Ibérica y Canarias: mecanismos de transporte y variación estacional, *3rd Asamblea Hispano portuguesa de Geodesia y Geofísica, Inst. Nac. de Meteorol.* Esp., Valencia
[3] Escudero M, Castillo S, Querol X, Avila A, Alarcón M, Viana MM, Alastuey A, Cuevas E and Rodríguez S 2005 Wet and dry African dust episodes over eastern Spain, *J. Geophys. Res.* 110, D18S08, doi:1029/2004JD004731
[4] Cachorro V E, Toledano C, de Frutos A M, Sorribas M, Vilaplana J M, and de la Morena B, 2005 Aerosol characterization at El Arenosillo (Huelva, Spain) with an AERONET/PHOTONS CIMEL sunphotometer *Geophys. Res. Abstract* **7**
[5] Obregón MA, Serrano A, Cancillo ML, Cachorro VE and Toledano C 2014 Aerosol radiometric properties at Western Spain (Cáceres station), *Int. J. Climatol.* DOI: 10.1002/joc.4031.
[6] Santos D, Costa M J and Silva A M 2008 Direct SW aerosol radiative forcing over Portugal, *Atmos. Chem. Phys.* **8** 5771-86
[7] Boucher O, Randall D, Artaxo P, Bretherton C, Feingold G, Forster P, Kerminen V M, Kondo Y, Liao H, Lohmann U, Rasch P, Satheesh S K, Sherwood S, Stevens B and Zhang Z Y 2013. Clouds and aerosols in climate change 2013. In: Stocker, T.F., Qin, D., Plattner, G.-K., Tignor, M., Allen, S.K., Boschung, J., Nauels, A., Xia, Y., Bex, V., Midgley, P.M. (Eds.), *The Physical Science Basis. Contribution of Working Group I to the Fifth Assessment Report of the Intergovernmental Panel on Climate Change.* Cambridge University Press, Cambridge, United Kingdom and New York, NY, USA
[8] Holben B, Eck T F, Slutsker I, Tanre D, Buis J, Setzer K, Vermote E, Reagan J, Kaufman Y, Nakajima T, Lavenu F, Jankowiak I and Smirnov A 1998 AERONET-A Federated Instrument Network and Data Archive for Aerosol Characterization. *Remote Sens. Environ.*, **66** 1-16
[9] Draxler R R, and Rolph GD, HYSPLIT (HYbrid Single-Particle Lagrangian Integrated Trajectory) Model access via NOAA ARL READY Website (http://www.arl.noaa.gov/HYSPLIT.php). NOAA Air Resources Laboratory, College Park, MD
[10] Rolph G D Real-time Environmental Applications and Display sYstem (READY) Website (http://www.ready.noaa.gov). NOAA Air Resources Laboratory, College Park, MD
[11] Rose T, Crewell S, Lohnert U, and Simmer C 2005 A network suitable microwave radiometer for operational monitoring of the cloudy atmosphere *Atmos. Res.* **75** 183–200

[12] Mayer B and Kylling A 2005 Technical note: The libRadtran software package for radiative transfer calculations – description and examples of use *Atmos. Chem. Phys.* **5** 1855–77

[13] Gueymard C 2004 The sun's total and spectral irradiance for solar energy applications and solar radiation models. *Sol. Energy* **76** 423–53

[14] Stamnes K, Tsay S C, Wiscombe W and Laszlo I 2000 DISORT, a general purpose Fortran program for discrete-ordinate-method radiative transfer in scattering and emitting layered media: documentation of methodology. Goddard Space Flight Center, NASA

Retrieval of the aerosol optical thickness from UV global irradiance measurements

M J Costa[1], V Salgueiro[2], D Bortoli[2], M A Obregón[2], M Antón[2,3] and A M Silva[2]

[1]Instituto de Ciências da Terra - Polo de Évora, Departamento de Física, Escola de Ciências e Tecnologia, Universidade de Évora, Évora, Portugal
[2]Instituto de Ciências da Terra - Polo de Évora, Departamento de Física, Instituto de Investigação e Formação Avançada, Universidade de Évora, Évora, Portugal
[3]Departamento de Física, Universidad de Extremadura, Badajoz, Spain

E-mail: mjcosta@uevora.pt

Abstract. The UV irradiance is measured at Évora since several years, where a CIMEL sunphotometer integrated in AERONET is also installed. In the present work, measurements of UVA (315 - 400 nm) irradiances taken with Kipp&Zonen radiometers, as well as satellite data of ozone total column values, are used in combination with radiative transfer calculations, to estimate the aerosol optical thickness (AOT) in the UV. The retrieved UV AOT in Évora is compared with AERONET AOT (at 340 and 380 nm) and a fairly good agreement is found with a root mean square error of 0.05 (normalized root mean square error of 8.3%) and a mean absolute error of 0.04 (mean percentage error of 2.9 %). The methodology is then used to estimate the UV AOT in Sines, an industrialized site on the Atlantic western coast, where the UV irradiance is monitored since 2013 but no aerosol information is available.

1. Introduction

The strong influence of aerosols in atmospheric processes and consequently in climate makes their quantification in terms of the aerosol optical thickness (AOT) to be extremely important [1]. The AERONET (AErosol RObotic NETwork) program [2-3] represents an important step towards a long-term database of aerosol optical, microphysical and radiative properties, providing globally distributed observations of these quantities in diverse aerosol regimes.

Although ultraviolet (UV) radiation accounts for less than 10 % of the solar radiation, the UV irradiance reaching the surface has particular significance within the solar spectrum due to its potential harmful effects, constituting a hazard for several life forms on Earth. For this reason UV radiation is being monitored since several years in regions with high insolation values [4]. In the absence of clouds, aerosols are the major regulators of the Earth radiation budget and play a central role in modulating UV radiation. Therefore, UV irradiance measurements may be used to estimate the atmospheric aerosol load under cloud-free conditions, constituting a useful tool to estimate AOT values in regions where only the UV radiation is monitored but no aerosol information is available, as was already proposed for pyranometer data [5].

In this work, the UV irradiance measured at Évora since several years is combined with radiative transfer calculations to retrieve the AOT in the UVA spectral region. The simultaneous existence of a CIMEL sunphotometer integrated in AERONET allows for checking the validity of the methodology applied, enabling its application to different sites. This work is organized as follows: after the introduction, a brief description of the sites and instrumentation used is presented in section 2; the methodology is presented in section 3 followed by the results and discussion in section 4. Finally, the conclusions are summarized in section 5.

2. Sampling sites and instrumentation

Simultaneous measurements of the UV radiation and aerosol characteristics are taken in Évora, at the Atmospheric Physics Observatory of the University of Évora (Institute of Earth Sciences - ICT) (38°34' N, 7°54' W, 293 m a.s.l). The site is located about 100 km eastward from the Atlantic western coast and it is in the pathway of different air masses, experiencing the influence of long-range transport of anthropogenic aerosols emitted in northern Europe, and by desert dust plumes advected from Africa. Desert dust and smoke from forest fires (often in summer), account for a significant amount of the suspended particle mass [6-9]. The UV irradiance is also monitored at the University of Évora pole in Sines, a small town located on the Portuguese western Atlantic coast, about 150 km south of Lisbon (37°57' N, 8°51' W, 15 m a.s.l), characterized by the existence of a fairly large industrial complex, one of the largest in the country. It is therefore of great interest the monitoring of the AOT at this site where air pollution is a major problem for the local population.

Measurements of the UVA (315 - 400 nm) irradiances are being taken with Kipp & Zonen radiometers at Évora (since 2011) and Sines (since 2013). The broadband radiometers are periodically calibrated at the Atmospheric Sounding Station "El Arenosillo" of the National Institute for Aerospace Technology in South-Western Spain, similarly to the erythemal broadband radiometers [10-11]. The Évora site has also a CIMEL CE-318 sunphotometer and is integrated in AERONET [2]. This sunphotometer provides measurements of the direct solar radiation with a 1.2° full field of view at 340, 380, 440, 500, 675, 870, 940 and 1020 nm. In addition, measurements of sky radiances in the almucantar and principal planes geometries, at 440, 675, 870 and 1020 nm, are also performed by the instrument. The AERONET AOT at 340 and 380 nm are used here to evaluate the accuracy of the AOT retrieved from the UVA irradiance measurements.

3. Methodology

The UVA irradiance is simulated using the 1D radiative transfer package LibRadtran (Library for Radiative Transfer) [12] considering the spectral response function of the radiometer (see figure 1), for varying conditions of solar zenith angle, total ozone column and AOT as shown in table 1 and a fixed surface albedo of 0.035. This allows for obtaining LookUp Tables (LUTs) that will subsequently be used to obtain the AOT from UVA irradiance measurements.

Table 1. Solar zenith angle, total ozone columns and aerosol optical thickness used in the radiative transfer simulations.

Quantity	Values
Solar zenith angle (°)	0 to 70, step of 1
Total ozone column (DU)	200 to 500, step of 5
Aerosol optical thickness	0.00001 0.01 0.03 0.06 0.1 0.15 0.2 0.25 0.3 0.4 0.5 0.6 0.8 1.0

A mid-latitude type of atmospheric vertical profile for winter or summer is considered in the calculations, according to the season [12]; aerosols are assumed to be described by the default properties [12], with rural aerosol type in the boundary layer, background aerosols above 2 km, spring-summer or fall-winter conditions (according to the season) and 50 km visibility. The use of a fixed aerosol type may introduce some bias under situations of African dust advection or smoke from forest fires, when the particle load is high, nevertheless, in the majority of the situations the approximation proposed is quantitatively accurate [13].

The radiative transfer equation is numerically solved using the discrete ordinates method with 16 quadrature angles. The extraterrestrial solar spectral irradiance is taken from the SORCE satellite mission that provides measurements of incoming x-ray, UV, VIS, NIR, and total solar radiation, available online at http://lasp.colorado.edu/home/sorce/data/ssi-data/ (last accessed 07/06/2015). The SORCE Solar Spectral Irradiance (SSI) composite data used here is constructed using measurements from the SOLSTICE (115-310 nm; spectral resolution: 1 nm) and SIM (310-2400 nm; spectral resolution: 1 to 27 nm) instruments, for 2013. The use of SORCE SSI is justified by the comparisons shown in Figure 1, between the SORCE SSI used here and one of the SSI contained in LibRadtran. Radiative transfer calculations in the UV spectral region are especially sensitive to the solar irradiance considered, since the smaller values with respect to the visible would translate into more substantial errors in the AOTs. Considerable differences between both can be distinguished in the graph of figure 1, which would lead to significant differences in the LUTs and consequently also in the derived AOTs.

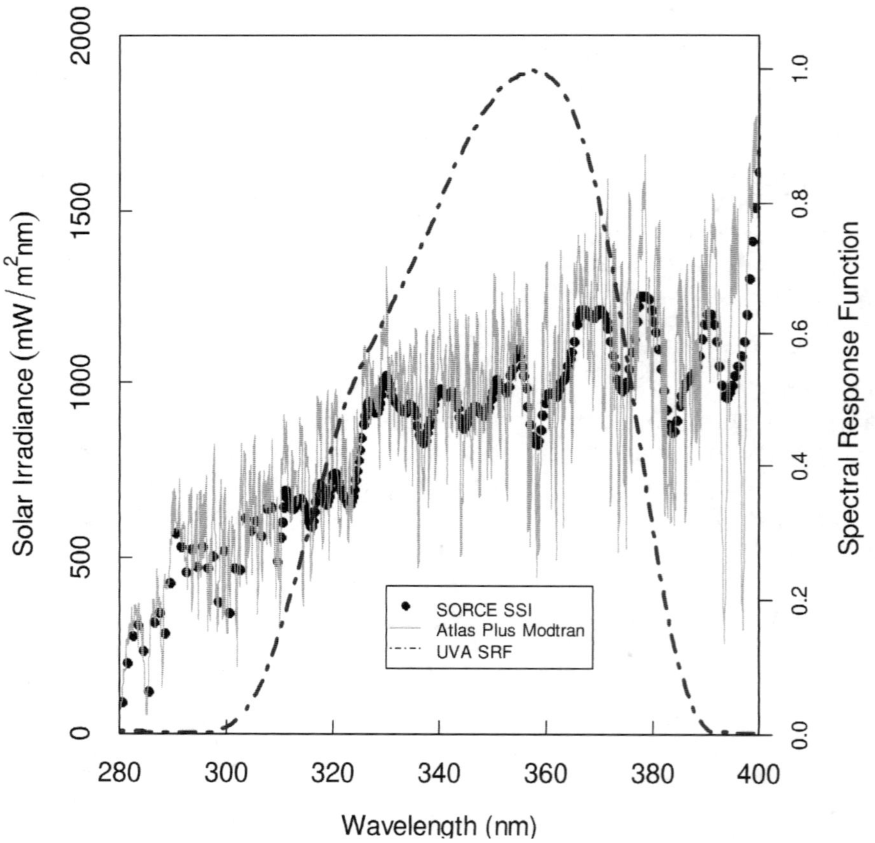

Figure 1. Spectral solar irradiance (SSI) in the UV region. Black dots represent the SORCE satellite data, whereas the gray line shows one of the SSI data contained in LibRadtran (Atlas Plus Modtran). The blue dash-dotted line represents the spectral response function of the UVA radiometer used.

The AOT values are then obtained by simple interpolation of the measured UVA irradiance, with a temporal resolution of 2 minutes, using the modelled irradiances contained in the LUTs. For each UVA measurement, the corresponding LUT (matching solar zenith angle and total ozone columns) is used, yielding the corresponding interpolated AOT value. The daily total ozone column is obtained from the Ozone Monitoring Instrument (OMI) flying onboard Aura satellite (ftp://toms.gsfc.nasa.gov/pub/omi/data/overpass/; last accessed 07/06/2015). Cloud screening is done through the inspection of the daily global UVA irradiance. Comparisons between the retrieved AOT results and corresponding AERONET AOT level 2.0 data, at 340 and 380 nm, are then carried out.

4. Results and discussion

The evolution of the AOT values retrieved from the UVA irradiance measurements in Évora is presented for each month of 2013 in the graphs of figure 2. From January to March, as well as from October to December the AOT values are in general lower than during spring and summer. The absence of data during June is connected to the participation of the UVA radiometer in a calibration campaign. In general the UVA derived AOT agree quite well with AERONET data, accompanying their behavior during background and event situations.

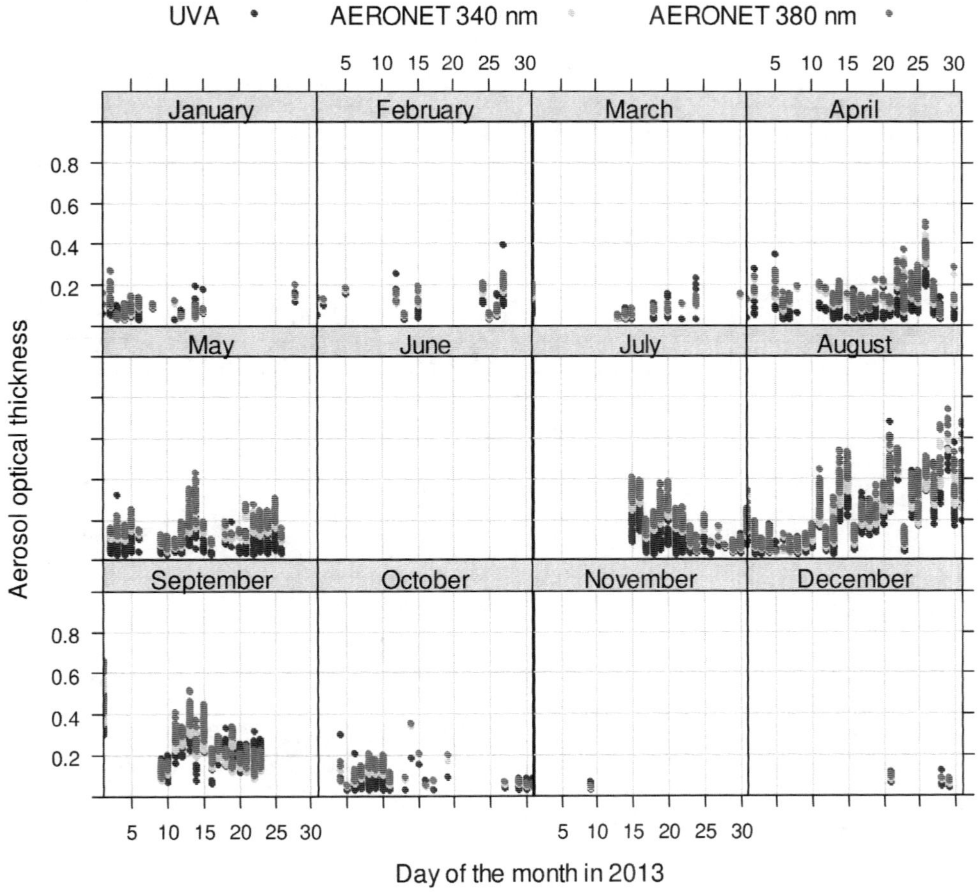

Figure 2. Time series of the AOT in Évora during 2013. Blue dots represent the AOT derived from the UVA irradiance measurements, whereas green and red dots represent the AERONET AOT at 340 and 380 nm, respectively.

The agreement between AERONET AOT and the UVA AOT retrieved for Évora during 2013 is investigated through the scatter plots shown in figure 3. It can be noted that the derived UVA AOT slightly underestimates AERONET AOT, being the agreement with AERONET AOT at 380 nm somewhat better. Curiously the month with the best correlation is August, when the highest AOT values are reached, which seems to show that the method performs better for situations with high aerosol load.

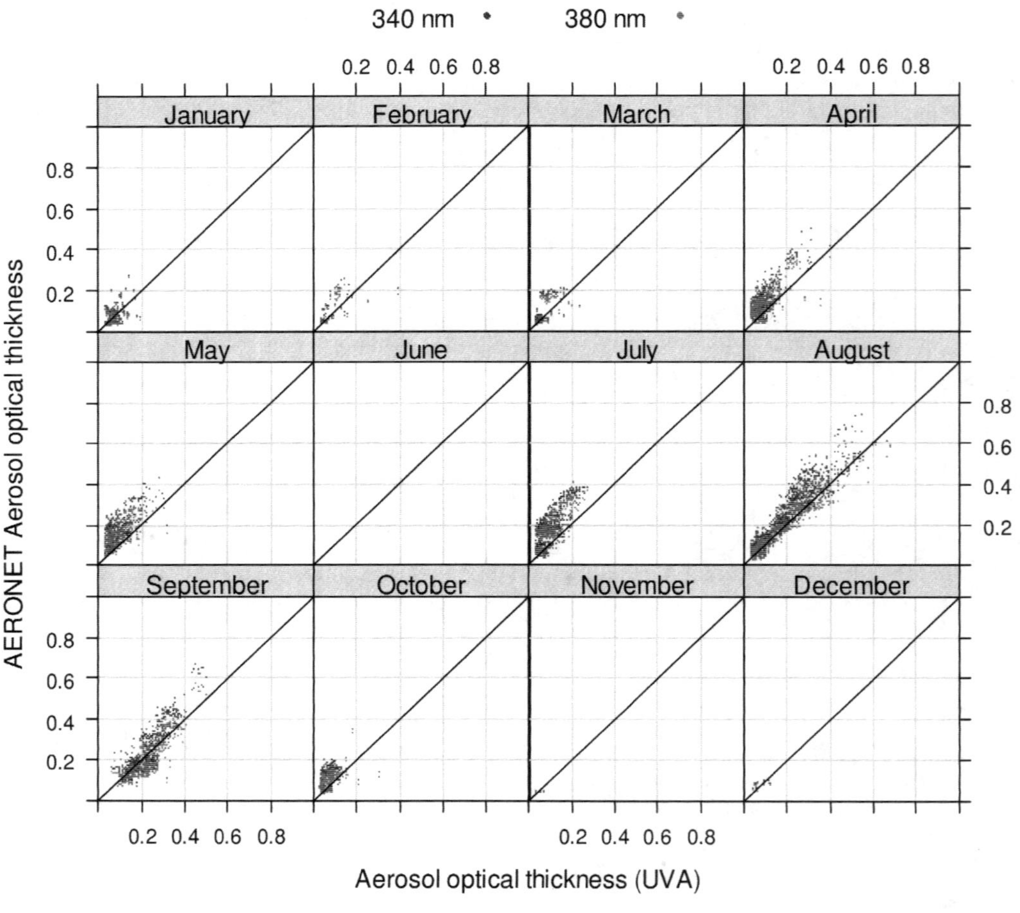

Figure 3. Scatter plots of the AERONET AOT versus the UVA AOT retrieved, for year 2013. The blue and green dots represent the AERONET AOT at 340 and 380 nm, respectively.

Figure 4 shows the smoothed scatter plot of the AERONET AOT at 380 nm versus the UVA AOT retrieved, for year 2013 in Évora. The linear regression confirms the tendency of the retrieved values to slightly underestimate the AOT obtained from AERONET at 380 nm, showing a slope of 0.862.

The correlation coefficient is rather high (R=0.86) and the root mean square error (RMSE) is found to be 0.05 (NRMSE of 8.3%) and the mean absolute error (MAE) of 0.04 (MPE of 2.9%). Overall results show a fairly good agreement between both AOT datasets, therefore the method may be useful in sites where the UVA irradiance is monitored but no aerosol information is accessible.

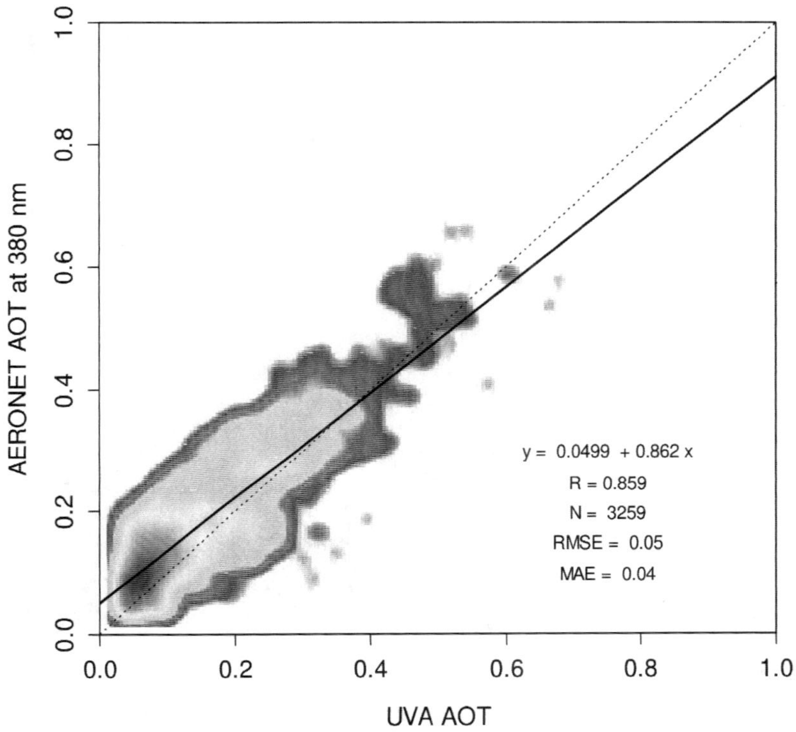

Figure 4. Smoothed scatter plot of the AERONET AOT at 380 nm versus the UVA AOT retrieved, for year 2013 in Évora. Color gradient from magenta to dark redish indicates an increasing density of points. The solid line represents the linear regression performed and the dashed line the 1:1 relation.

Hence, the methodology is subsequently applied to the UVA irradiance data of the Sines site, where no aerosol information is available, providing an estimate of the aerosol load in the atmosphere. The dataset used corresponds to the spring-summer of 2014 (June to September), since in these seasons south-western Iberia is often affected by high atmospheric aerosol loads originated from the transport of desert dust favoured by the meteorological conditions, or from forest fires. In addition, as aforementioned Sines hosts a large industrial area that influences air quality in the area, probably contributing to increase the aerosol load. Marine aerosols also have here an important contribution since this is a costal site. The hourly averaged AOT results obtained when applying the methodology to the UVA irradiance measured in Sines are presented in figure 5. The results obtained for Évora in the same period are also presented for comparisons purposes. The AOT in Sines is systematically higher than in Évora and this is most probably connected with the air pollution issues in the industrialized site of Sines. As for Évora, a background site with no industries around, the main origin of aerosol events is the long range transport of desert dust and smoke from forest fires. The results also suggest that Sines is also affected by long range transport of aerosols, since the events that affect Évora in mid June and in the end of July, are also perceptible in the AOT of Sines. The fact that a fixed aerosol type is considered in this approach may introduce some uncertainties in the retrievals. In this regard, the higher AOT values obtained for Sines should be compared with independent measurements of the AOT, as for example CIMEL CE-318 sunphotometer retrievals, in order to assess the accuracy. This could be done in a near future intensive measurement campaign.

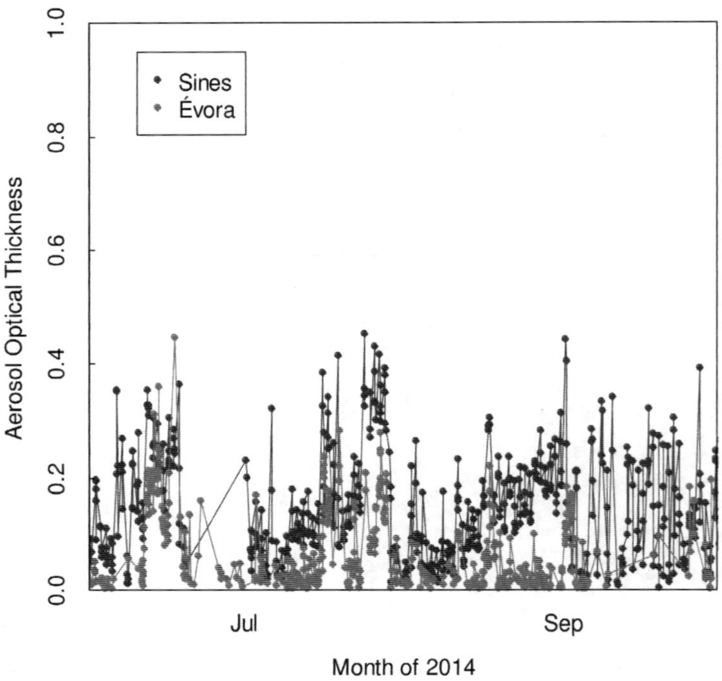

Figure 5. Time series of the hourly AOT derived from the UVA irradiance measurements in Sines and in Évora, during June – September 2014.

5. Conclusions

The work presents a methodology based on the combination of UV global irradiance data with radiative transfer calculations, to retrieve the AOT in the UVA spectral region. It is applied to data measured in Évora and the UVA AOT results are compared with AERONET AOT at 340 and 380 nm. The comparison shows that the derived UVA AOT slightly underestimates AERONET AOT, being the agreement with AERONET AOT at 380 nm slightly better. A high correlation coefficient is obtained for the comparisons (R=0.859). The RMSE is found to be 0.05 (NRMSE=8.3 %) and a MAE of 0.04 (mean percentage error of 2.9 %) is obtained. Overall results show a fairly good agreement between both AOT datasets, therefore the method is considered adequate to apply to different sites, whenever the UVA irradiance is monitored but no aerosol information is available.

Acknowledgments

This work was partially supported by FCT (Fundação para a Ciência e a Tecnologia) through grants SFRH/BD/88669/2012 and SFRH/BPD/86498/2012, and project FCOMP-01-0124-FEDER-014024 (PTDC/AAC-CLI/114031/2009) and also by the research project CGL2011-29921-C02-01/CLI granted by the "Ministerio de Economía y Competitividad" of Spain. The authors acknowledge the funding provided by ICT, under contract with FCT (the Portuguese Science and Technology Foundation).. The authors also acknowledge Samuel Bárias for maintaining instrumentation used in this work. Thanks are due to AERONET/PHOTONS and RIMA networks for the scientific and technical support. CIMEL calibration was performed at the AERONET-EUROPE GOA calibration center, supported by ACTRIS under agreement no. 262254 granted by European Union FP7/2007–2013.

References

[1] Boucher O *et al.* (Eds.), *The Physical Science Basis. Contribution of Working Group I to the Fifth Assessment Report of the Intergovernmental Panel on Climate Change.* Cambridge University Press, Cambridge, United Kingdom and New York, NY, USA

[2] Holben B *et al.* 1998 AERONET-A Federated Instrument Network and Data Archive for Aerosol Characterization *Remote Sens. Environ.* **66** 1-16

[3] Holben B N, *et al.* 2001 An emerging ground-based aerosol climatology: Aerosol optical depth from AERONET *J. Geophys. Res.* **106** 12067-97

[4] Serrano A, Antón M, Cancillo M L and Mateos V L 2006 Daily and annual variations of erythemal ultraviolet radiation in Southwestern Spain *Ann. Geophys.* **24** 427-41

[5] Lindfors A V, Kouremeti N, Arola A, Kazadzis S, Bais A F and Laaksonen A 2013 Effective aerosol optical depth from pyranometer measurements of surface solar radiation (global radiation) at Thessaloniki, Greece *Atmos. Chem. Phys.* **13** 3733-41

[6] Obregón M A , Pereira S, Salgueiro V, Costa M J, Silva A M , Serrano A and Bortoli D 2015 Aerosol radiative effects during two desert dust events in August 2012 over the Southwestern Iberian Peninsula Atmos. Res. **153** 404-15

[7] Santos D, Costa M J and Silva A M 2008 Direct SW aerosol radiative forcing over Portugal *Atmos. Chem. Phys.* **8** 5771-86

[8] Pereira S, Wagner F and Silva A M 2008 Scattering properties and mass concentration of local and long-range transported aerosols over the South Western Iberia Peninsula *Atmos. Environ.* **42** 7623-31

[9] Elias T, Silva A M, Belo N, Pereira S, Formenti P, Helas G and Wagner F 2006 Aerosol extinction in a remote continental region of the Iberian Peninsula during summer *J. Geophys. Res.* **111** 1-20

[10] Vilaplana J M *et al.* 2009 *Report of the El Arenosillo/INTA-COST calibration an intercomparison campaign of UVER broadband radiometers*, Huelva, Spain, Aug.-Sept. 2007, http://www-med-physik.vu-wien.ac.at/uv/ ISBN: 978-84-692-2640 7, COST Office, p 64

[11] Antón M, Serrano A, Cancilllo M L and Vilaplana J M 2011 Quality assurance of broadband erythemal radiometers at the Extremadura UV Monitoring Network (Southwestern Spain), *Atmos. Res.* **100** 83-92

[12] Mayer B and Kylling A 2005 Technical note: The libRadtran software package for radiative transfer calculations – description and examples of use *Atmos. Chem. Phys.* **5** 1855–77

[13] Salgueiro V, Costa M J, Silva A M and Bortoli D 2014 Variability of the Daily-Mean Shortwave Cloud Radiative Forcing at the Surface at a Midlatitude Site in Southwestern Europe *J. Climate* **27** 7769-80

IOP Conf. Series: Earth and Environmental Science **28** (2015) 012008 doi:10.1088/1755-1315/28/1/012008

Determination of selected polycyclic aromatic compounds in particulate matter: a validation study of an agitation extraction method for samples with low mass loadings using reduced volumes

S. García-Alonso[1], R.M. Pérez-Pastor[1], V. Archilla-Prat[2], J. Rodríguez-Maroto[1], M. Izquierdo-Díaz[2], E. Rojas[1] and D. Sanz[1]

[1]CIEMAT, Av. Complutense, 22, Madrid, 28040, Spain

[2]Turbojet Engine Test Centre, INTA, Ctra Ajalvir km4, Torrejón de Ardoz, Madrid, 28850, Spain

E-mail: susana.garcia@ciemat.es

Abstract

A simple analytical method using low volumes of solvent for determining selected PAHs and NPAHs in PM samples is presented. The proposed extraction method was compared with pressurized fluid (PFE) and microwave (MC) extraction techniques and intermediate precision associated to analytical measurements were estimated. Extraction by agitation with 8 mL of dichloromethane yielded recoveries above 80% compared to those obtained from PFE extraction. Regarding intermediate precision results, values between 10-20% were reached showing increases of dispersion for compounds with high volatility and low levels of concentration. Within the framework of the INTA/CIEMAT research agreement for the PM characterization in gas turbine exhaust, the method was applied for analysis of aluminum foil substrates and quartz filters with mass loading ranged from 0.02 to 2 mg per sample.

1. Introduction

The analysis of polycyclic aromatic compounds (PACs) usually involves High Performance Liquid Chromatography with fluorescence detector (HPLC/FD) which is well known because of its good selectivity and high sensitivity [1]. Regarding classical sample-preparation techniques, soxhlet (SX), microwave solvent (MC), pressurized fluid extraction (PFE) and ultrasonic extractions (US) are mostly used. However, important disadvantages associated includes time-consuming/large amount of solvent for SX, expensive for PFE and MC, and lower extraction efficiency for US. Current trends in analytical chemistry have focused to simplify sample preparation and minimize solvent volume, so efforts are directing towards miniaturization of extraction procedures by reducing the amount of sample mass and volume of organic solvent.

Because of this, it is very important to develop analytical protocols which assesse quality of results. Additionally, the validity of results has to be demonstrated. Regarding method validation, the best option would be to participate in inter-comparison exercises for the compounds in the selected matrix to determine the accuracy and precision of methodology. When it is not possible, the use of multiple

methods could be recommended as a validation to obtain comparable data, because of the advantages and disadvantages of different analytical methods give additional credence to the results.
This work presents the results obtained for validation of a simple sample treatment procedure to analyze selected PACs by agitation extraction using reduced volumes of solvent. The main focus was the analysis of PACs in particulate matter with low mass loading, such as impactor substrates used to determine the nano and submicron size distribution.

2. Experimental

2.1. Sample collection
The sampling site was the National Institute for Aerospace Technology (INTA), situated to 20 km Northeast of Madrid city. The site is an open urban area with direct influence from airport sources.
A high volume MCV CAV-Amb sampler loaded with Whatman glass fibre filters (102 mm diameter) was used for the sampling of the particulate matter (PM10). Ambient air samples were taken during 24 – 48 hours (30 m^3 h^{-1}) for consecutive days of November and June (2014). The Samples then were divided into four sub-samples which were weighted and then considered as independent sub-samples for analysis.

2.2. Sample preparation and chromatographic analysis
- Microwave Solvent Extraction (MC): PM sub-samples were extracted using an Ethos Sel (Millestone) at 900 W (120ºC) and dichloromethane as solvent (15 mL).
- Pressurised Fluid Extraction (PFE): PM sub-samples were extracted using an ASE350 (Dionex Corporation) and dichloromethane (100 ºC, 10.34 psi and 1 static cycles). The total volume of extract corresponded to 55 mL.
- Sonication/agitation (AGITATION): PM sub-samples were extracted in closed tubes containing dichloromethane (8 mL), sonicated (15 min) and shaken for short-time. Re-extraction was done to assure recovery and the combined extracts were then treated as follows.

The extracts were then filtered (0.22 μm pore size PTFE) and slowly concentrated under nitrogen flow to change solvent (methanol, 150-200 μL) for HPLC/FLD analysis.
Once selected PAHs were determined, derivatisation reaction of extracts was performed for NPAH determination [2]. 0.5 mL aqueous solution of $CuCl_2$ (0.05%) and 0.05 g of $NaBH_4$ were added to methanol extract. 1 mL of dichloromethane was added and then shaken for extraction of amino-PAHs. Extract was concentrated under nitrogen flow to change solvent (acetonitrile, 100 μL).

3. Results

3.1. Comparison of results among microwave, pressurized extraction and agitation extraction
The detection limit was considered as twice the standard deviation on average of replicate analyses of 3 standard solutions (1 ng mL^{-1}). Experimental values varied between 0.002-0.003 and 0.15 ng $injection^{-1}$ for investigated PAHs and NPAHs, respectively. More volatile PAHs, such as methyl naphthalene, acenaphthene and fluorene also corresponded to 0.15 ng $injection^{-1}$ as detection limits.
Five ambient air samples (about 40 mg of mass aerosol loading/filter) were divided as 4 parallel sub-samples. The first sub-sample was extracted by MC, the second one by PFE and the third by AGITATION, using dichloromethane as extraction solvent. The comparison was performed by grouping results corresponding to extraction technique and deducing mean and standard deviation for normalised values (figure 1). Results were similar among extraction methods, reaching recoveries above 80% for AGITATION if compared with those of PFE extraction.

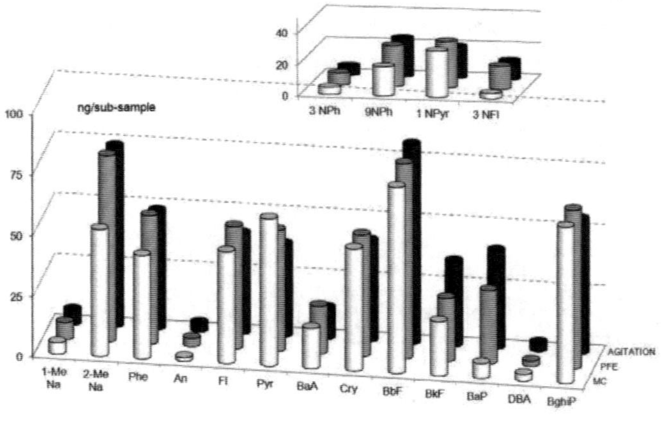

Figure 1. Mean values (ng sub-sample^{-1}) of selected PACs measured using microwave (MC), pressurised fluid (PFE) and sonication/agitation (AGITATION).

Methyl-naphthalene (Me-Na), phenanthrene (Ph), anthracene (An), fluoranthene (Fl), pyrene (Pyr), benz[a]anthracene (BaA), chrysene (Cry), benz[b]fluoranthene (BbF), benz[k]fluoranthene (BkF), benz[a]pyrene (BaP), dibenz(a,h)anthracene (DBA) and benzo[g,h,i]perylene BghiP

3.2. Estimation of the intermediate precision associated to results from real sample analyses

- MC: Four samples were divided into four sub-samples and analysed by pairs to compare microwave and agitation procedures. Sub-sample mass loading was 5 - 8 mg of collected particulate matter.
- PFE: Analogously, three ambient air filters were divided and analysed by pairs to compare PFE/AGITATION. Sub-sample mass loading was 0.7 - 2.8 mg of collected particulate matter.

Once all data obtained, intermediate precision estimations of measurements was performed based on pooled precision (RSD_{pool})

The term corresponds to the intermediate precision of the method and has been considered as the standard deviation values deduced from duplicate analysis of each air sample using both techniques. In particular, we have considered the relative standard deviations (RSD) of the results of each ambient air sample in duplicate (fv_n), grouping and deducing the value of pooled relative standard deviation (RSD_{pool}). For comparison between the two techniques of extraction, the calculation was separately carried out in both cases using the following equation [3]:

$$RSD_{pool} = \sqrt{\frac{\sum RSD_{fv_n}^2 (n_{fv_n} - 1)}{\sum (n_{fv_n} - 1)}}$$

From the results of agitation procedure, as can be seen, the RSD associated to PAH analysis was generally below 10 % for concentration levels above 10 ng sub-sample^{-1} (for mass loading of sample: 5-8 mg). Higher variability was associated (10-15%) when concentration levels were between 1 and 5 ng sub-sample^{-1} (for mass loading of sample: 0.7-2.8 mg).

Table 1. Values of intermediate precision deduced from results by analysing samples using MC and agitation (mass loading of sample 5-8 mg) and those for PFE and agitation procedures (mass loading of sample 0.7-2.8 mg). Estimations were based on pooled standard deviations (RSD pool).

	mass loading of sample: 5-8 mg		mass loading of sample : 0.7-2.8 mg	
	MC	AGITATION	PFE	AGITATION
1-Me Na	0.45	0.27		
2-Me Na	0.48	0.23		
Ace	0.15	0.20		
F	0.12	0.15		
Phe	0.19	0.14	0.36	0.28
An	0.20	0.15		
Fl	0.07	0.06	0.15	0.10
Pyr	0.07	0.14	0.12	0.08
BaA	0.08	0.06	0.15	0.10
Cry	0.09	0.07	0.18	0.15
BbF	0.06	0.06	0.16	0.12
BkF	0.06	0.06	0.16	0.11
BaP	0.21	0.07	0.11	0.09
BghiP	0.08	0.07	0.17	0.11
3 NPh	0.29	0.26	0.27	0.43
9 NPh	0.17	0.11	0.20	0.36
1 NPyr	0.12	0.27	0.16	
3NFl	0.29	0.22		

For microwave results, an important reduction in the recovery and increases of variability (up to 50 %) were found for methyl naphthalene, the most volatile. These observations likely correspond to losses due to the heat applied and the more volatile nature of these compounds.
For NPAHs, intermediate deviation values were higher, around 20 %, which is justified by the increase of experimental variability associated with additional derivatization step for chromatographic analysis. Calculations could be performed when concentration levels were well measured (mass loading 5-8 mg), but in the case of low concentrations (mass loading 0.7-2.8 mg) no enough data were obtained for 3NFl (PFE and agitation) and 1 NPyr (agitation). PFE shows a slightly lower variability of the results regarding the agitation extraction.

3.3. Analytical application

Several aerosol samples were taken on quartz and aluminium substrates (by means of a Low Pressure Impactor and a holder filter) during the preliminary studies on characterization from the turbofan engine exhaust [4]. Mass loadings greater than 2 mg per sample were collected on quartz filters and exceeding to 20 µg per sample on aluminium foil substrates.
In total, 16 sampling substrates were analysed and maximum/mean concentration values were below 10 ng substrate^{-1} as it is compiled in table 5. No selected NPAHs were detected in this kind of samples.

Table 2. Concentration range (ng substrate^{-1}) measured from quartz and aluminum substrates.

	mass loading> 2mg	mass loading> 20μg
ng substract^{-1}	quartz	aluminum
1-Me Na	<0.4-2.3	<0.4-9.6
2-Me Na	<1.0-5.2	<1.0-39
Ace	<0.4-1.7	<0.4-0.6
F	<0.2-2.3	<0.2-1.3
Phe	<0.7-16	<0.7-9.5
An	<0.03-1.2	<0.03-0.3
Fl	<0.01-11	<0.01-1
Pyr	<0.01-18	<0.01-1.8
BaA	<0.01-5.5	<0.01-0.04
Cry	<0.05-11	<0.05
BbF	<0.01-6.4	<0.01-0.2
BkF	<0.004-9.2	<0.004-0.1
BaP	<0.01-9.2	<0.01-0.3
DBA	<0.01-2	<0.01
BghiP	<0.02-5.8	<0.02-2.2

4. Conclusions

It has been developed an analytical method for the determination of certain PACs in PM samples collected on filters by sonication/agitation and subsequent analysis by HPLC/FD. The method is simple and requires the use of low volumes of solvent for extraction.
Due to the lack of availability of reference materials at these low mass loadings, we chose to analyse real PM samples divided into four aliquots which were tested by agitation, microwave and pressurized fluid extractions for comparison. An estimation of the variability of results has been also conducted for intermediate precision calculations. Finally, the effectiveness of the proposed method was evaluated by analysing aerosol samples collected on quartz and aluminum substrates in extremely low concentration range, such as impactor substrates used to determine the nano and submicron size distribution.

5. References

[1] OSPAR Commission 2003 *JAMP guidelines on methods and criteria for harmonised sampling and analysis of PAHs in air and precipitation* (Ref. No. 2003-8)
[2] Barrado-Olmedo AI, Pérez-Pastor RM, and García-Alonso S 2012 *Talanta* **101** 428
[3] Pantazopoulos P 2001 *Uncertainty Estimate for: Procedure ONT-F CL-0024 "Determination of Ochratoxin A in Wine and Grape Juice by High Performance Liquid Chromatography with Fluorescence Detection* (Food Lab. Div. Ontario Region)
[4] Rodríguez-Maroto JJ, Archilla V, Sanz-Rivera D, Rojas-García E, Pujadas M, Mosquera JL, Mercader D, Gonzalez A, Jimenez A, Entero A, Mena J, Fernández-Mainez JM and Bezares JC 2013 *RICTA 2013* (Évora, Portugal)

IOP Conf. Series: Earth and Environmental Science **28** (2015) 012009 doi:10.1088/1755-1315/28/1/012009

Effect of water vapor in the SW and LW downward irradiance at the surface during a day with low aerosol load

M A Obregón[1], M J Costa[2], A Serrano[3] and A M Silva[2]

[1] Instituto de Ciências da Terra - Polo de Évora, Departamento de Física, Instituto de Investigação e Formação Avançada, Universidade de Évora, Évora, Portugal
[2] Instituto de Ciências da Terra - Polo de Évora, Departamento de Física, Escola de Ciências e Tecnologia, Universidade de Évora, Évora, Portugal
[3]Departamento de Física, Universidad de Extremadura, Badajoz, Spain

E-mail: nines@unex.es

Abstract. The aim of this study is to analyze the effect of the water vapor content on the downward irradiance measured at the Earth's surface. For that purpose, downward irradiance values have been estimated with the radiative transfer model libRadtran in different spectral ranges: shortwave (SW: 285-2800 nm) and longwave (LW: 3500-50000 nm), and with different water vapor content in the column. These simulations have been made for Évora, Portugal, the August 4, 2012, a cloud-free day and with low aerosol optical depth.
The comparison between the simulated irradiance with different water vapor contents shows differences in both spectral ranges. For SW, the irradiance reaching the surface increases when the water vapor content decreases, obtaining an increase of up to 4 %, 2 % and 1%, corresponding the largest increases to the smallest values of water vapor. For LW, the behaviour is the opposite, the irradiance decreases when the water vapor content decreases, obtaining a decrease of up to 10 %, 4 % and 2 %, corresponding the largest decreases to the smallest values of water vapor. The effect of water vapor in the aerosol radiative forcing (ARF) has also been analyzed, obtaining relative difference values of up to 2.5 % for SW and 35 % for LW.

1. Introduction

Water vapor is an important atmospheric component, influencing the Earth's climate in many ways. It plays a key role in the global radiation budget, in energy transport mechanisms in the atmosphere as well as in photochemical processes. As for the radiative budget, the water vapor is a key element, contributing directly by means of infrared radiation absorption emitted by the Earth's surface and the atmosphere, or indirectly by means of microphysical processes favoring the formation of clouds, and affecting the size, shape and chemical composition of aerosols and thus modifying the aerosol role in the radiative forcing.

Estimations of downward irradiance provided by a reliable radiative transfer code, as libRadtran model [1], using different water vapor contents in the column are of great interest in order to analyze the water vapor effects in the irradiance and in the radiative balance of the Climate System. Therefore, the aim of this study is to analyze the effect of different water vapor contents in the simulated downward irradiance values at the surface in different spectral ranges: shortwave and longwave. This work is organized as follows: a brief description of the study region and instrumentation is presented

IOP Conf. Series: Earth and Environmental Science **28** (2015) 012009 doi:10.1088/1755-1315/28/1/012009

in section 2; data set and methodology are provided in section 3; results are discussed in section 4. Finally, conclusions are given in section 5.

2. Study region and instrumentation

The location of Évora radiometric station is shown in Figure 1. It is installed in the Évora pole of the Institute of Earth Sciences in Évora, whose geographical coordinates are: 38.6° N, 7.9° W, 293.0 m a.s.l. This station is located near the center of a small town with about 60,000 inhabitants, about 100 km eastward from the Atlantic west coast. Évora is influenced by different aerosol types, namely urban as well as mineral and forest fire aerosol particles [2-7].

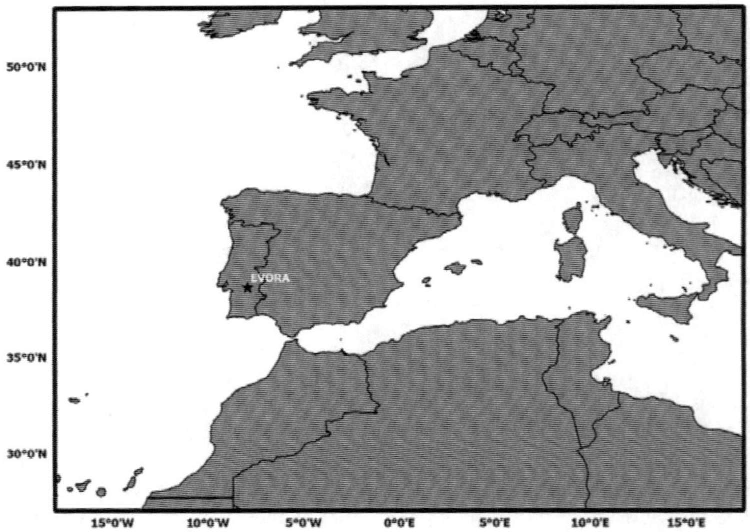

Figure 1. Iberian Peninsula showing the location of Évora station.

Évora station is managed by the Évora pole of the Institute of Earth Sciences, at the University of Évora (Portugal). This station is equipped with an Eppley Black & White pyranometer, an Eppley pyrgeometer and CIMEL CE-318 sunphotometers, among several other radiometric instruments. An Eppley Black & White pyranometer measures the global shortwave irradiance (285-2800nm), providing 10 minutes averages of 10 seconds sampling time. The uncertainty associated with this instrument is estimated to be about 5% encompassing calibration, temperature and cosine characteristics of the radiometer. The Eppley pyrgeometer is a broad-band infrared radiometer of wide spectral range (λ> 4000 nm), and measures the downward flux density of atmospheric radiation at the horizontal surface with a temporal sampling of 10 minutes. The CIMEL CE-318 sunphotometer is integrated in the NASA AERONET (Aerosol Robotic NETwork) network [8], make direct sun measurements with a 1.2° full field of view at 340, 380, 440, 500, 675, 870, 940 and 1020 nm. In addition, measurements of sky radiances in the almucantar and principal planes geometries, at 440, 675, 870 and 1020 nm, are also performed by this instrument. The channel of 940 nm channel is used to estimate the precipitable water vapor [9]. This instrument is operating continuously in Évora since 2003. More details about this instrument are given by Holben et al. (1998) [8]. The parameters obtained from Cimel sunphotomers and used in this study are aerosol optical depths (AOD), Ångström α exponent (440-870) (α), single scattering albedo (ω), asymmetry factor (g) and precipitable water vapor column (PWC). These parameters have been used as input to the radiative transfer model.

3. Dataset and methodology

In this study, the effect of different water vapor content on the downward irradiance values at the surface in different spectral ranges, shortwave (SW: 285-2800 nm) and longwave (LW: 3500-50000 nm), has been analyzed. This was done through the comparison of 10-min averaged estimations of downward irradiance using different water vapor contents, provided by libRadtran model [1]. These estimations have been made for Évora, Portugal, the August 4, 2012, a cloud-free day and with low aerosol optical depth.

The inputs of libRadtran model are: AERONET level 2.0 aerosol optical properties, AERONET precipitable water vapor column (PWC) and surface albedo, as well as total ozone column provided by the Ozone Monitoring Instrument (OMI). The values of the total column water vapor used in the simulations are: true value (corresponding to AERONET PWC), true value – 2, true value – 5 and true value – 10. The total column water vapor for that day ranges between 18.1 and 20.2 mm, being the average value equal to 19.4 mm. The level 2.0 aerosol properties used as input were: α and β Ångström coefficients (α is obtained here for the 440 and 870 nm wavelength range and the turbidity β from α and value and aerosol optical depth, AOD, at 1020 nm). Also, the asymmetry parameter (the average value of this parameter for the four wavelengths available (440, 675, 870 and 1020 nm) was used). The level 2.0 AERONET single scattering albedo, $\omega(\lambda)$, dataset is quite scarce since only cases with AOD at 440 nm greater than 0.4 are considered, which is very unusual in Évora. Therefore, a value of aerosol single scattering albedo equal to 0.95 for all wavelengths was used in model calculations of irradiance. The value of 0.95 is the average value in the level 2.0 in Évora and it is consistent with previous observations reported for this station [7]. The surface albedo values considered are those described in Obregón et al. (2015) [10], for the same station. The other variables taken into account in setting up the model simulations are the following: the extraterrestrial irradiance values (obtained from Gueymard (2004) [11]), profiles of temperature, air density, ozone and other atmospheric gases (taken from the midlatitude summer/winter standard atmospheres) and the radiative transfer equation solver (the discrete ordinate method of Stamnes et al. (2000) [12], DISORT2 with 16 streams, was used). To obtain the 10 minutes input values, interpolations from the existing AERONET properties values were made.

The water vapor can modify the aerosol role in the radiative forcing, and therefore the effect of water vapor in the aerosol radiative forcing (ARF) has also been analyzed. ARF is the effect of atmospheric aerosol particles on the radiation balance at a given level and indicates the magnitude of change in the radiative balance due to changes in aerosol physical/optical properties. The ARF is defined here as the difference between the downward global solar irradiance at the surface in the presence of aerosols, I_d, and the same quantity in background/baseline conditions, I^0_d, (Eq. 1), assuming that, in background/baseline conditions, the aerosol optical depth is null, as:

$$ARF = I_d - I_d^0 \, . \qquad \text{Eq. 1}$$

In addition, SW and LW radiations were measured by an Eppley pyranometer and an Eppley pyrgeometer, installed at the Évora Geophysics Center Observatory in Évora. Only cloud-free measurements corresponding to solar zenith angle lower than 80° have been considered in this study.

4. Results and discussion

The aim of this section is to analyze and to compare the effect of different water vapor contents on downward irradiance at the surface in both spectral ranges, SW and LW. Figures 2 and 3 show the temporal evolution of 10-min averaged downward irradiance values simulated with different water vapor contents, for SW and LW, respectively. It is shown that there are notable differences in both spectral ranges. For SW, the irradiance reaching the surface increases when the water vapor content decreases. For LW, the behaviour is opposite, the irradiance decreases when the water vapor content also decreases. This behaviour is also seen in Figure 4, where relative differences between the simulated irradiances with lower water vapor content respect to the simulated irradiances with the true water vapor content are shown. These relative differences indicate that SW irradiance increase up to 4 %, 2 % and 1%, respectively, corresponding the largest increases to the smallest values of water vapor, and the LW irradiance decrease up to 10 %, 4 % and 2 %, respectively, corresponding the largest decreases to the smallest contents of water vapor.

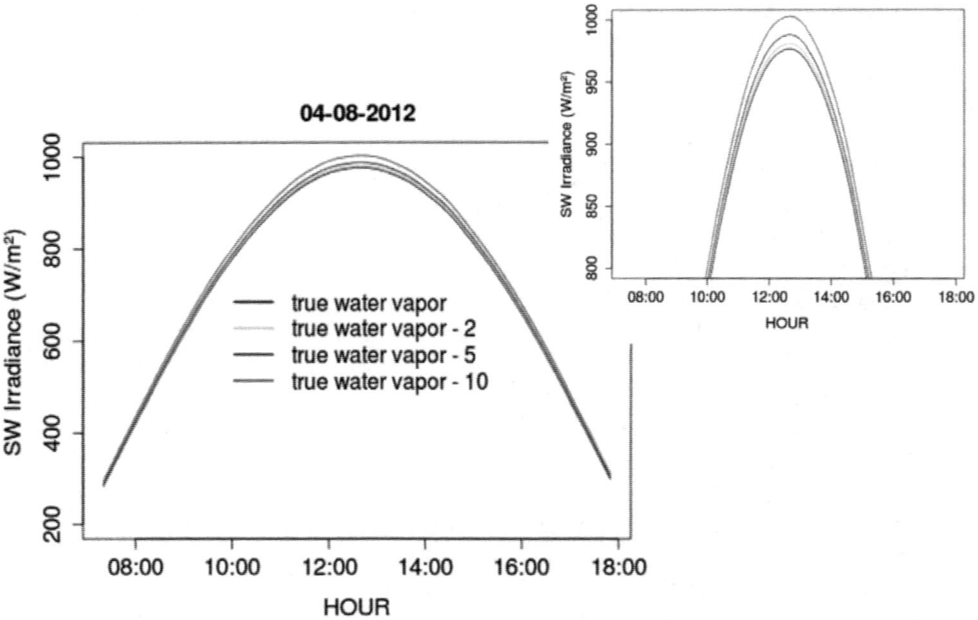

Figure 2. Temporal evolution of simulated SW irradiance with different water vapor contents.

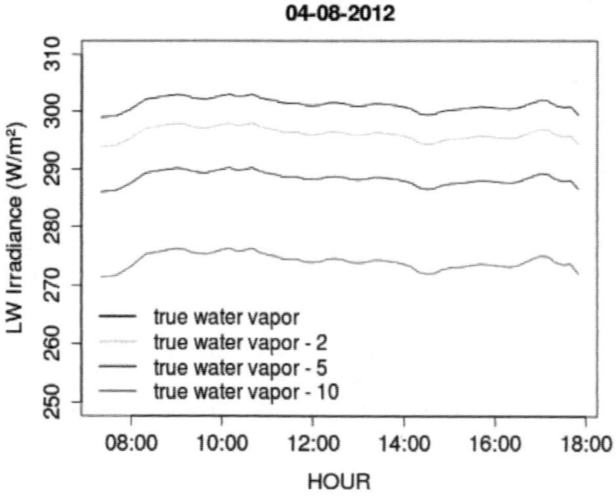

Figure 3. Temporal evolution of simulated LW irradiance with different water vapor contents.

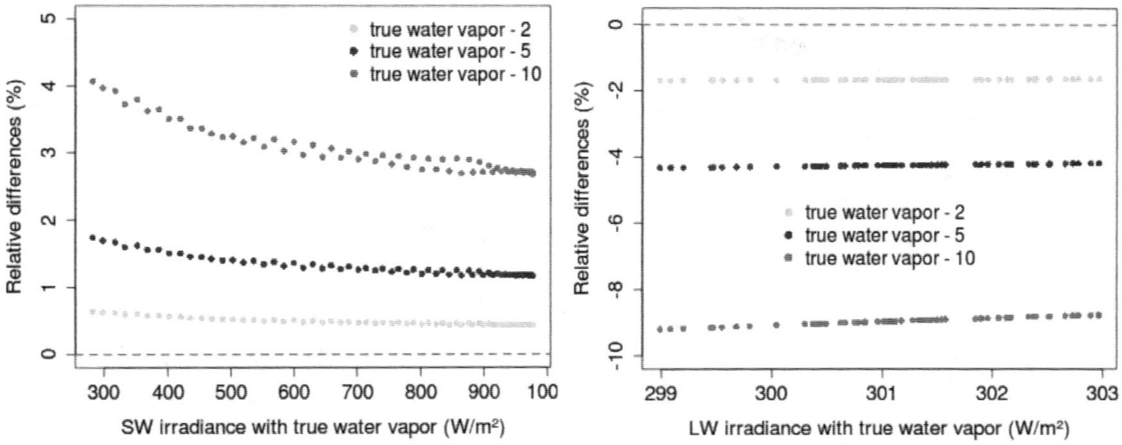

Figure 4. Relative differences between the simulated irradiances with true water vapour -2, true water vapour -5 and water vapor -10, with respect to the simulated irradiances with the true water vapor content. The figure on the left corresponds to SW and the figure on the right corresponds to LW.

The effect of water vapor in the ARF has also been analyzed, obtaining relative differences values, between the ARF calculated with lower water vapor content respect to the ARF calculated with the true water vapor content, of up to 2.5 % for SW and 35 % for LW (figure 5). These values also indicate that the water vapor affects to SW and LW irradiances, but this effect is greater in LW. ARF values in LW may vary about 35% when the true value of total water vapor -10 is used.

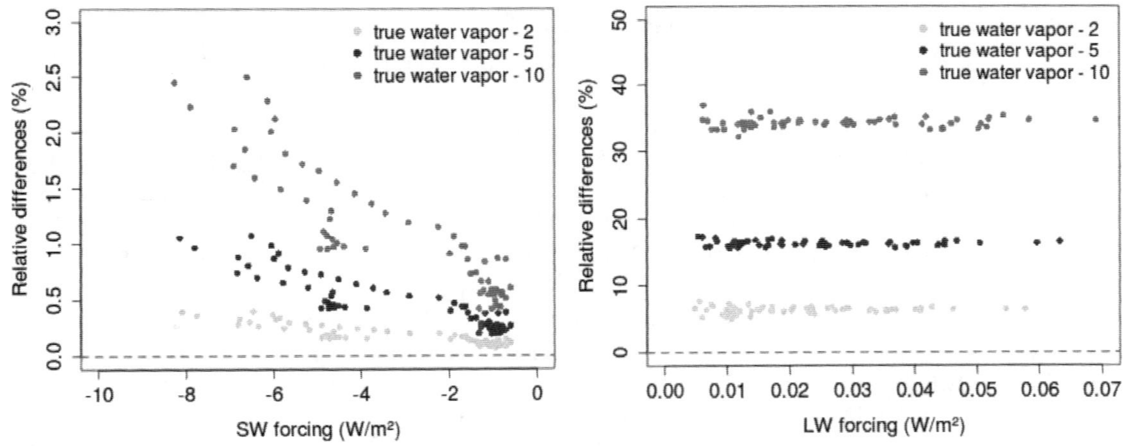

Figure 5. Relative differences between the ARF calculated with true water vapour -2, true water vapour -5 and water vapor -10, with respect to the ARF calculated with the true water vapor content. The figure on the left corresponds to SW and the figure on the right corresponds to LW.

5. Conclusions

This study contributes to the study of the effect of different water vapor contents on downward irradiance at the surface in different spectral ranges, shortwave and longwave The effect of the water vapor was analyzed through the comparison between 10-min averaged values of downward irradiance at the surface simulated with the libRadtran radiative transfer code and from different values of the total water vapor content in the column. From this comparison it can be concluded that the SW irradiance reaching the surface increases when the water vapor content decreases, whereas for LW the behaviour is opposite, i.e., the irradiance decreases when the water vapor content decreases.

This study shows the significant influence of water vapor content on the SW and LW irradiance measured at the Earth's surface, being more important in the LW. Therefore, it is concluded that, in order to obtain accurate estimations of the irradiance, the model must be fed with highly reliable values of the water vapor content.

Acknowledgments
This work was partially supported by FCT (Fundação para a Ciência e a Tecnologia) through the grant SFRH/BPD/86498/2012, the project PDTC/CEO-MET/4222/2012 and the research project CGL2011-29921-C02-01/CLI granted by the "Ministerio de Economía y Competitividad" of Spain. The authors acknowledge the funding provided by ICT, under contract with FCT (the Portuguese Science and Technology Foundation). The authors also acknowledge Samuel Bárias for maintaining instrumentation used in this work. Thanks are due to AERONET/PHOTONS and RIMA networks for the scientific and technical support. CIMEL calibration was performed at the AERONET-EUROPE GOA calibration center, supported by ACTRIS under agreement no. 262254 granted by European Union FP7/2007–2013.

References

[1] Mayer B and Kylling A 2005 Technical note: The libRadtran software package for radiative transfer calculations – description and examples of use Atmos. Chem.Phys., 5, 1855–1877
[2] Pereira S, Silva A M, Elias T and Wagner F 2005 Aerosol monitoring at Cabo da Roca site", paper presented at 4º Simpósio de Meteorologia e Geofísica da APMG/6º Encontro Luso-Espanhol de Meteorologia, Sesimbra, Portugal, 2005.
[3] Pereira S, Wagner F and Silva A M 2008 Scattering properties and mass concentration of local and long-range transported aerosols over the South Western Iberia Peninsula Atmos. Environ. 42 (33), pp. 7623-7631, 2008, http://dx.doi.org/10.1016/j.atmosenv.2008.06.008.
[4] Pereira S, Wagner F and Silva A M 2011 Seven years of measurements of aerosol scattering properties, near the surface, in the south-western Iberia Peninsula Atmos. Chem. Phys., vol.11, pp.17-29, 2011. http://dx.doi.org/10.5194/acp-11-17-2011.
[5] Elias T, Silva A M, Belo N, Pereira S, Formenti P, Helas G and Wagner F 2006 Aerosol extinction in a remote continental region of the Iberian Peninsula during summer J. Geophys. Res. 111 (D14204), 1-20, 2006. http://dx.doi.org/10.1029/2005JD006610.
[6] Silva A M, Wagner F, Pereira S and Elias T 2006 Aerosol properties at the most western point of continental Europe, presented at International Aerosol Conference, Minnesota, USA.
[7] Obregón M A, Pereira S, Wagner F, Serrano A, Cancillo M L and Silva A M 2012 Regional differences of column aerosol parameters in western Iberian Peninsula. Atmos. Environ. 12, 1–10. http://dx.doi.org/10.1016/j.Atmosenv.2012.08.016.
[8] Holben B, Eck T F, Slutsker I, Tanre D, Buis J, Setzer K, Vermote E, Reagan J, Kaufman Y, Nakajima T, Lavenu F, Jankowiak I and Smirnov A 1998 AERONET-A Federated Instrument Network and Data Archive for Aerosol Characterization. Remote Sens. Environ.,vol. 66, pp.1–16,1998.
[9] Bruegge C J, Conel J E, Green R O, Margolis J S, Holm R G and Toon G 1992 Water vapor abundance retrievals during FIFE, J. Geophys. Res., 97(D17), 18,759 – 18,768
[10] Obregón M A , Pereira S, Salgueiro V, Costa M J, Silva A M , Serrano A and Bortoli D 2015 Aerosol radiative effects during two desert dust events in August 2012 over the Southwestern Iberian Peninsula. Atmos. Res. 153, 404–415. http://dx.doi.org/10.1016/j.atmosres.2014.10.007.
[11] Gueymard C 2004 The sun's total and spectral irradiance for solar energy applications and solar radiation models. Sol. Energy 76, 423–453.
[12] Stamnes K, Tsay S C, Wiscombe W and Laszlo I 2000 DISORT, a general purpose Fortran program for discrete-ordinate-method radiative transfer in scattering and emitting layered media: documentation of methodology. Goddard Space Flight Center, NASA.

IOP Conf. Series: Earth and Environmental Science **28** (2015) 012010 doi:10.1088/1755-1315/28/1/012010

Validation of libRadtran and SBDART models under different aerosol conditions

M A Obregón[1], A Serrano[2], M J Costa[3] and A M Silva[3]

[1] Instituto de Ciências da Terra - Polo de Évora, Departamento de Física, Instituto de Investigação e Formação Avançada, Universidade de Évora, Évora, Portugal
[2]Departamento de Física, Universidad de Extremadura, Badajoz, Spain
[3] Instituto de Ciências da Terra - Polo de Évora, Departamento de Física, Escola de Ciências e Tecnologia, Universidade de Évora, Évora, Portugal

E-mail: nines@unex.es

Abstract. This study is aimed at analyzing the performance of the radiative transfer models libRadtran and SBDART for estimating the shortwave (285-2800 nm) irradiance under different aerosol conditions. For that purpose, measurements of downward irradiance at the Earth´s surface in Évora and the corresponding model simulations have been compared for six selected days (03-08-2003, 15-08-2004, 24-08-2004, 18-07-2005, 19-08-2007 and 17-05-2010). The comparison between measured and simulated values shows a highly significant correlation, with determination coefficient between 0.999 and 1 for both models and slope very close to unity (between 0.992±0.004 and 1.017±0.003 for libRadtran, and between 0.997±0.004 and 1.024±0.002 for SBDART), supporting the validity of the models in the estimation of irradiance in the shortwave spectral range. Relative differences between the simulated and measured irradiances with respect to the measured values have also been calculated, being most of the differences for the six days lower than 0.2 %. These small differences could be associated with experimental errors in the measurements as well as uncertainties in the input values given to the models, particularly those related with the aerosol properties. Also, relative differences between models have been calculated, obtaining values lower than 0.1 %. SBDART model slightly overestimates the libRadtran simulations. The notably good agreement between simulated and measured irradiances indicates that both models can be used to estimate irradiance, provided that the models are fed with highly reliable-quality data.

1. Introduction

It is well known the general interest for accurately quantifying the effects of atmospheric aerosols on the energy balance of the Earth-atmosphere system. Their role involves the attenuation of solar radiation through scattering and absorption processes, and also the modulation of terrestrial radiation by scattering, absorbing and emitting. Atmospheric aerosols also indirectly affect the radiation balance by influencing the cloud formation and their properties. According to the Intergovernmental Panel on Climate Change 2013, the total aerosol radiative forcing is estimated to be -0.9 [-1.9 to -0.1] W m^{-2} [1]. Therefore, a global cooling effect due to the aerosols is now relatively well established, despite the uncertainties still remaining. Thus, accurate and reliable measurements and scientific assessment are still the order of the day.

Estimations of irradiance provided by reliable radiative transfer codes are of great interest in order to analyze the aerosol effects in the radiative balance of the Climate System. They are also of great interest when no measurements are available. Therefore, the aim of this study is to analyze the performance of the radiative transfer models libRadtran [2] and SBDART [3] for estimating the shortwave irradiance under different aerosol conditions over Évora station, Portugal. This work is organized as follows: a brief description of the study region and instrumentation is presented in section 2; dataset and methodology are provided in section 3; results are discussed in section 4. Finally, conclusions are given in section 5.

2. Study region and instrumentation

Évora radiometric station is installed in the Geophysics Center Observatory in Évora, with geographical coordinates: 38.6º N, 7.9º W, 293 m a.s.l. This station is located near the center of a small town with about 60000 inhabitants, about 100 km eastward from the Atlantic west coast. Évora is influenced by different aerosol types, namely urban as well as mineral and forest fire aerosol particles [4-9]. Évora station is managed by the Évora pole of the Institute of Earth Sciences, at the University of Évora (Portugal).

This station is equipped with an Eppley Black & White pyranometer and CIMEL CE-318 sunphotometers, among several other radiometric instruments. The Eppley Black & White pyranometer measures the global shortwave irradiance (285-2800nm) every 10 seconds and records 10 minutes averages. The uncertainty associated with this instrument is estimated to be about 5% encompassing calibration, temperature and cosine characteristics of the radiometer. The CIMEL CE-318 sunphotometer is integrated in the NASA AERONET (Aerosol Robotic NETwork) network [10]. It takes direct sun measurements with a 1.2º full field of view at 340, 380, 440, 500, 675, 870, 940 and 1020 nm, and measures of sky radiances in the almucantar and principal planes geometries, at 440, 675, 870 and 1020 nm. More details about this instrument are given by [10]. In this study, aerosol optical depths (AOD), Ångström α exponent (440-870) (α), single scattering albedo (ω), asymmetry factor (g) and precipitable water vapor column (PWC) provided by Cimel sunphotometer are used as inputs to the radiative transfer models for simulating the irradiance in the shortwave spectral range and to analyze the aerosol concentration of each day.

3. Dataset and methodology

In this study the reliability of the version 1.7 of the libRadtran [2] and of the version 2.4 of the SBDART [3] models for estimating the shortwave irradiance during days with different aerosol concentrations has been analyzed. This was done through the comparison between 10-min averages of the measured downward irradiance at the surface and the corresponding model simulations, only it takes into account measurements and simulations with zenith angle lower to 80°. Previously, the aerosol properties over Évora station have been analyzed and six days with different aerosol concentration and without cloud have been selected. Most days correspond to summer, when it is easier to find cloudless days and with aerosol measurements.

The inputs are the same in the two models: AERONET level 2.0 aerosol optical properties, precipitable water vapor column (PWC) and surface albedo, as well as total ozone column provided by the Ozone Monitoring Instrument (OMI). The level 2.0 aerosol properties used as input were: α and β Ångström coefficients (α is obtained here for the 440 and 870 nm wavelength range and the turbidity β from α and value and aerosol optical depth, AOD, at 1020 nm), and the aerosol optical depth for all wavelengths. Also, the asymmetry parameter (the average value of this parameter for the four wavelengths available (440, 675, 870 and 1020 nm) was used). The level 2.0 AERONET single scattering albedo, ω(λ), dataset is quite limited because only cases where AOD at 440 nm is greater than 0.4 are considered, which is very unusual in Évora. Therefore, a typical value of aerosol single scattering albedo equal to 0.95 for all wavelengths was used in model calculations of irradiance. This value corresponds to the average value in the level 2.0 data of Évora and it is consistent with previous observations reported for this station [9]. The surface albedo considered is that described for the same station in Obregón et al. (2015) [11]. The other variables taken into account in setting up the model simulations are the following: the extraterrestrial irradiance values (obtained from Gueymard (2004) [12]), profiles of temperature, air density, ozone and other atmospheric gases (taken from the midlatitude summer/winter standard atmospheres) and the radiative transfer equation solver (the discrete ordinate method of Stamnes et al. (2000) [13], DISORT2 with 16 streams, was used). To obtain the 10 minutes input values, interpolations from the existing AERONET properties values were made.

In addition, radiation was measured by an Eppley pyranometer installed at the Évora Geophysics Center Observatory in Évora. Only cloud-free measurements corresponding to solar zenith angle lower than 80° have been considered in this study.

4. Results and discussion

The days selected in this study to analyze the performance of the models for estimating the shortwave irradiance are shown in Table 1. This table also shows the daily average values of AOD 440 nm and α during the six days. On 03-08-2003 and 24-08-2004 AOD present very high values, 0.34 and 0.39 respectively. Simultaneously, the values of α for these days are the smallest, 0.54 and 0.21 respectively, indicating the coarse mode predominance. These values indicate the influence of aerosol from desert areas during these days. During the others four days of the study, the concentration of aerosols is lower and the values of α is higher, indicating that particles are smaller. During these four days there is great variety of situations, because AOD values range between 0.03 and 0.14, and α values range between 0.94 and 1.83.

Table 1. Daily average values of AOD and α for the six days used in this study.

DATE	AOD 440 nm	α (440-870)
03-08-2003	0.34	0.54
15-08-2004	0.03	0.94
24-08-2004	0.39	0.21
18-07-2005	0.06	1.73
19-08-2007	0.14	1.83
17-05-2010	0.06	1.28

Figure 1 shows the comparison between the 10-min averaged downward irradiance measurements and simulated values. Measured and simulated values show a highly significant correlation, with determination coefficients between 0.999 and 1 for both models and slopes very close to unity (libRadtran between 0.992±0.004 and 1.017±0.003 and SBDART between 0.997±0.004 and 1.024±0.002). These values clearly support the validity of the models in the simulation of irradiances in the shortwave spectral range. This behaviour is also seen in Figure 2, where the temporal evolution of the 10-min averaged measurements and corresponding model simulations are shown.

Figure 3 shows the relative differences between the downward irradiance measurements and the corresponding simulated values for each day (red and blue dots). The relative differences between both models have been also represented (black dots). Relative differences between the simulated and measured irradiances with respect to the measured values indicate that most of the differences for the six days are lower than 0.2 %. These small differences could be associated with experimental errors in the measurements as well as with uncertainties in the input used in the model, particularly related with the actual aerosol properties. Also, relative differences between models have been calculated, obtaining values lower than 0.1 %. The SBDART model slightly overestimates the libRadtran simulations. The notably good agreement between simulated and measured irradiances indicates that both models can be used to estimate irradiance, provided that the models are fed with highly reliable data.

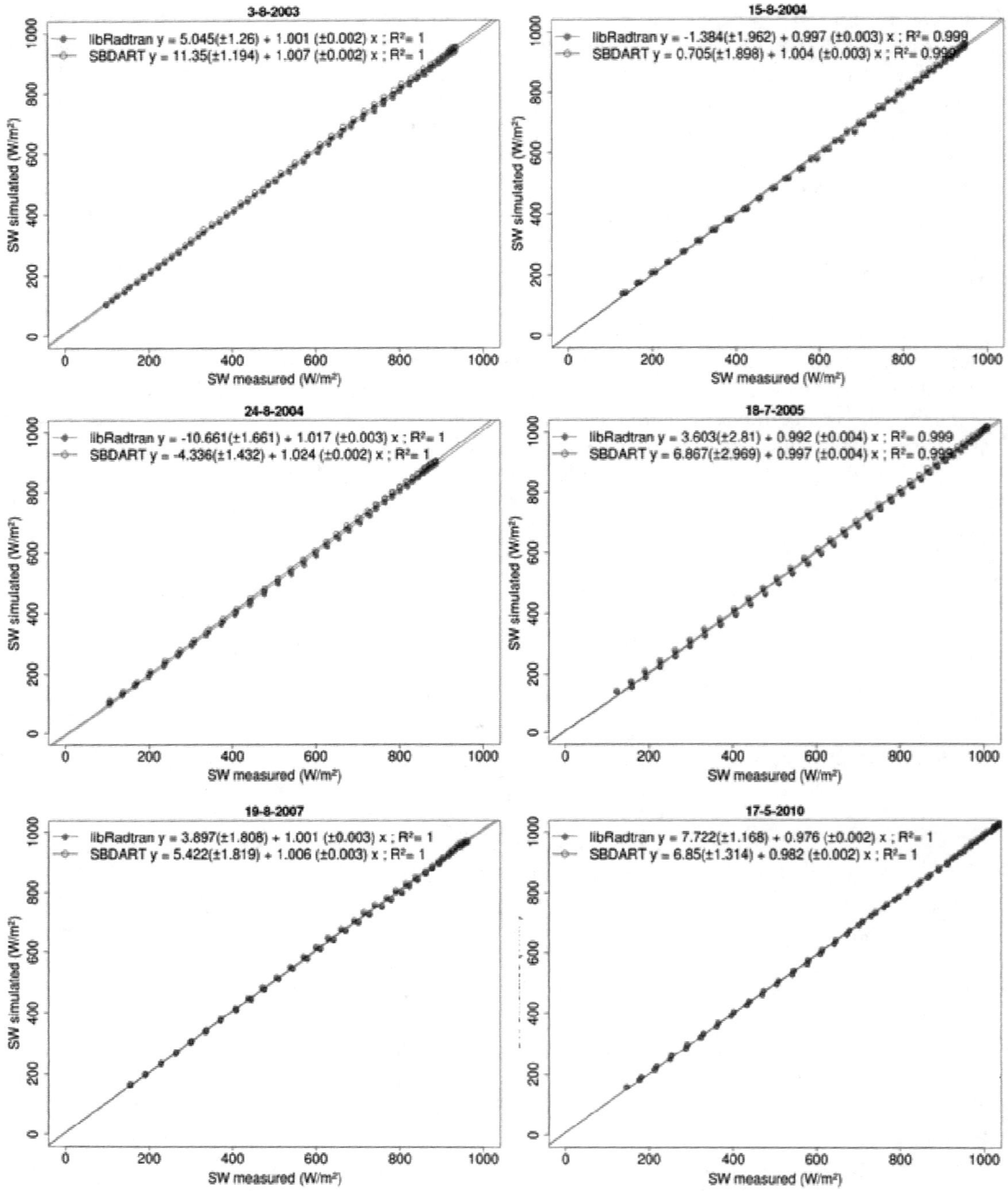

Figure 1. Comparison of simulated SW irradiances with the corresponding measurements. The solid lines represent the regression line for each model. The regression equations and determination coefficients are also included.

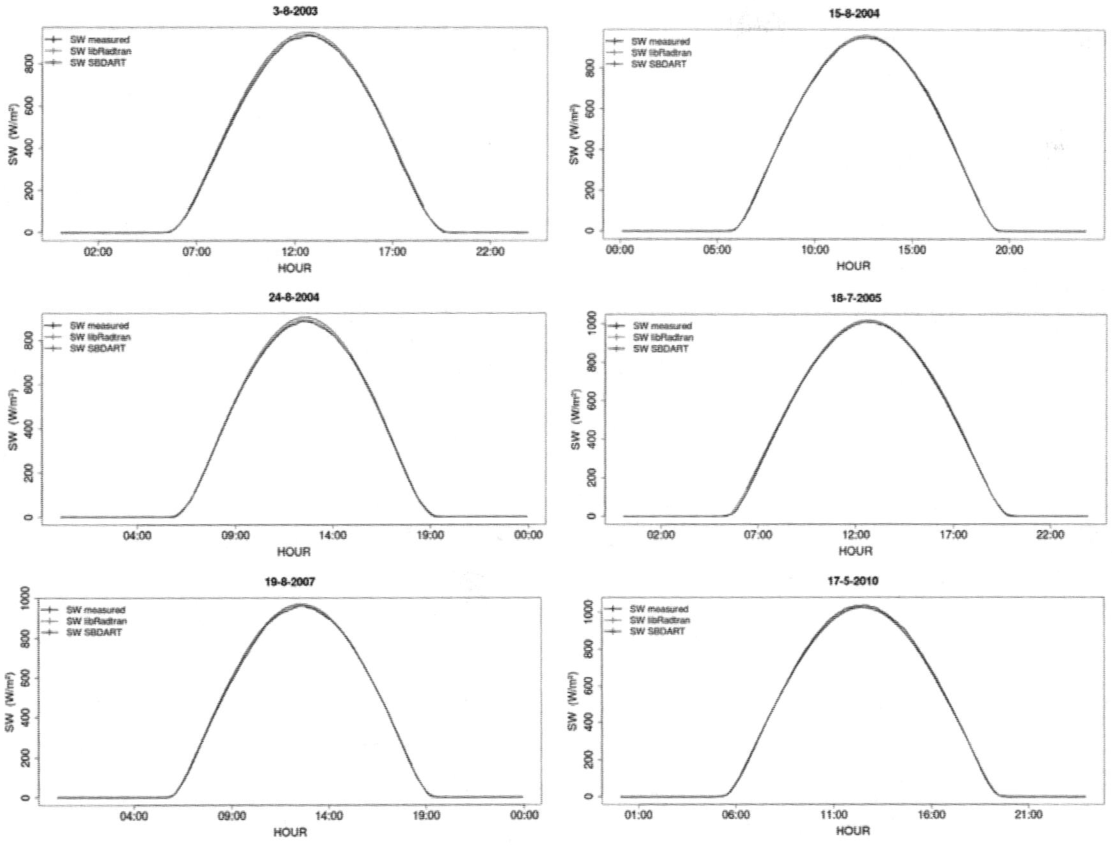

Figure 2. Temporal evolution of simulated and measured SW irradiances.

5. Conclusions

This study contributes to the estimation of downward irradiance, at the surface, in the shortwave spectral range, with libRadtran and SBDART models. The reliability of these models has been validated by the comparison between simulated and measured values observed over Évora under different aerosol concentrations. Determination coefficients between 0.999 and 1 for both models and slopes very close to unity (libRadtran between 0.992±0.004 and 1.017±0.003 and SBDART between 0.997±0.004 and 1.024±0.002) were obtained. Relative differences between the simulated and measured irradiances, with respect to the measured values, also confirm the reliability of the model, with most of the differences lower than 0.2 %. Also, relative differences between models have been calculated, resulting in values lower than 0.1 %.

The notably good agreement between simulated and measured irradiances indicates that both models can be used to estimate solar irradiance, provided that the models are fed with highly reliable data. These estimations can be used when no irradiance measurements are available and in future works to calculate aerosol radiative forcing values and analyze their effects in the radiative balance of the Climate System.

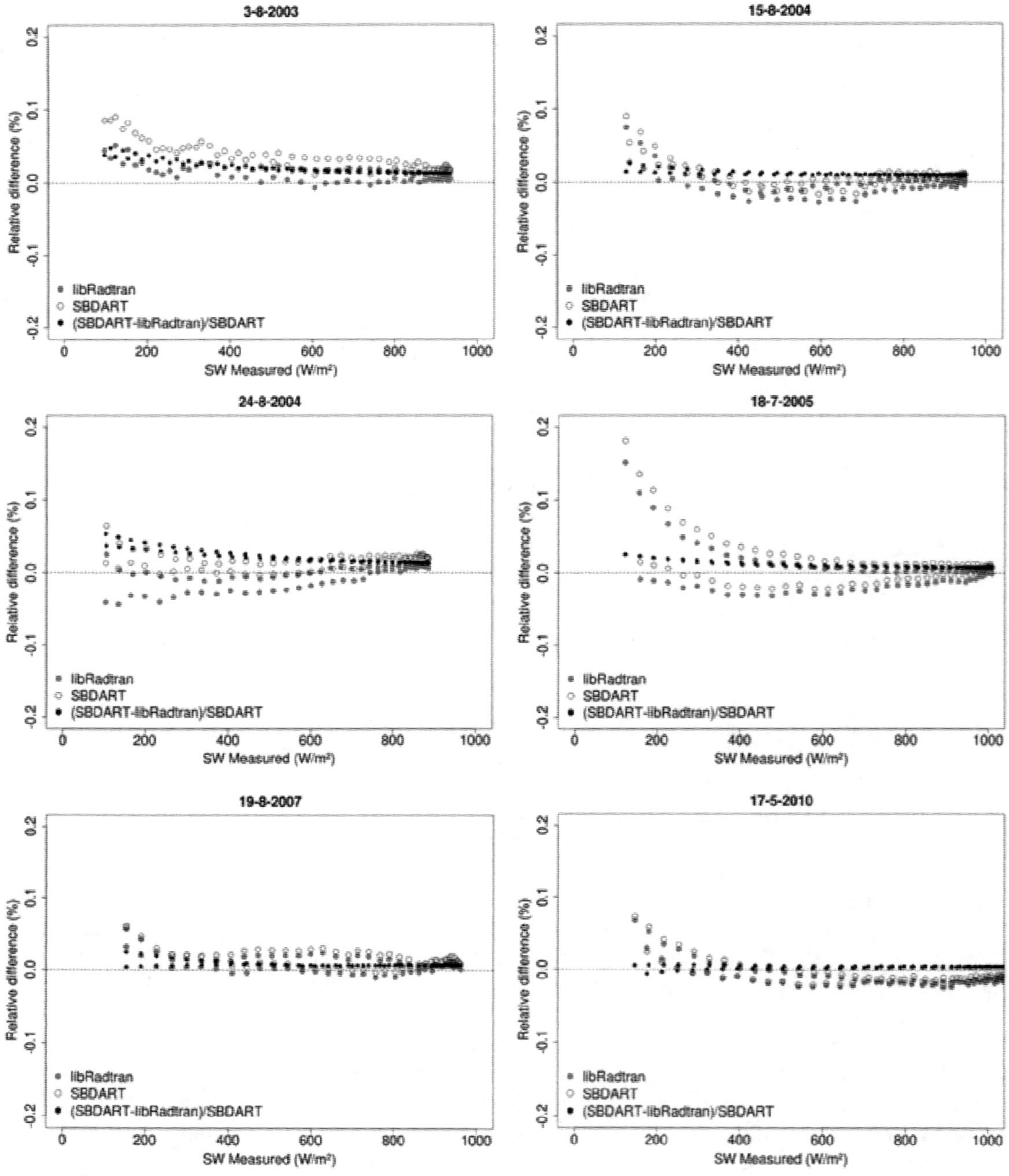

Figure 3. Relative differences between 10-min averaged measurements of downward irradiance at the surface and corresponding model simulations. The relative differences between models are also shown in these graphs (black filled dots).

Acknowledgments

This work was partially supported by FCT (Fundação para a Ciência e a Tecnologia) through the grant SFRH/BPD/86498/2012, the project PDTC/CEO-MET/4222/2012 and the research project CGL2011-29921-C02-01/CLI granted by the "Ministerio de Economía y Competitividad" of Spain. The authors acknowledge the funding provided by ICT, under contract with FCT (the Portuguese Science and Technology Foundation). The authors also acknowledge Samuel Bárias for maintaining instrumentation used in this work. Thanks are due to AERONET/PHOTONS and RIMA networks for the scientific and technical support. CIMEL calibration was performed at the AERONET-EUROPE GOA calibration center, supported by ACTRIS under agreement no. 262254 granted by European Union FP7/2007–2013.

References

[1] Boucher O et al. 2013. Clouds and aerosols in climate change 2013. In: Stocker, T.F., Qin, D., Plattner, G.-K., Tignor, M., Allen, S.K., Boschung, J., Nauels, A., Xia, Y., Bex, V., Midgley, P.M. (Eds.), *The Physical Science Basis. Contribution of Working Group I to the Fifth Assessment Report of the Intergovernmental Panel on Climate Change*. Cambridge University Press, Cambridge, United Kingdom and New York, NY, USA

[2] Mayer B and Kylling A 2005 Technical note: The libRadtran software package for radiative transfer calculations – description and examples of use *Atmos. Chem.Phys.* **5**, 1855–1877

[3] Ricchiazzi P, Yang S, Gautier C and Sowle D 1998, SBDART, A research and teaching tool for planeparellel radiative transfer in the Earth's atmosphere *B. Am. Meteorol. Soc.* **79**, 2101–2114

[4] Pereira S, Silva A M, Elias T and Wagner F 2005 Aerosol monitoring at Cabo da Roca site", paper presented at 4º Simpósio de Meteorologia e Geofísica da APMG/6º Encontro Luso-Espanhol de Meteorologia, Sesimbra, Portugal, 2005

[5] Pereira S, Wagner F and Silva A M 2008 Scattering properties and mass concentration of local and long-range transported aerosols over the South Western Iberia Peninsula *Atmos. Environ.* **42** (33), pp. 7623-7631, http://dx.doi.org/10.1016/j.atmosenv.2008.06.008

[6] Pereira S, Wagner F and Silva A M 2011 Seven years of measurements of aerosol scattering properties, near the surface, in the south-western Iberia Peninsula *Atmos. Chem. Phys.* **11**, pp.17-29, http://dx.doi.org/10.5194/acp-11-17-2011

[7] Elias T, Silva A M, Belo N, Pereira S, Formenti P, Helas G and Wagner F 2006 Aerosol extinction in a remote continental region of the Iberian Peninsula during summer *J. Geophys. Res.* 111 (**D**14204), 1-20, http://dx.doi.org/10.1029/2005JD006610

[8] Silva A M, Wagner F, Pereira S and Elias T 2006 Aerosol properties at the most western point of continental Europe, paper presented at International Aerosol Conference, Minnesota, USA

[9] Obregón M A, Pereira S, Wagner F, Serrano A, Cancillo M L and Silva A M 2012 Regional differences of column aerosol parameters in western Iberian Peninsula. *Atmos. Environ.* **12**, 1–10. http://dx.doi.org/10.1016/j.Atmosenv.2012.08.016

[10] Holben B, Eck T F, Slutsker I, Tanre D, Buis J, Setzer K, Vermote E, Reagan J, Kaufman Y, Nakajima T, Lavenu F, Jankowiak I and Smirnov A 1998 AERONET-A Federated Instrument Network and Data Archive for Aerosol Characterization. *Remote Sens. Environ.* **66**, pp.1–16

[11] Obregón M A , Pereira S, Salgueiro V, Costa M J, Silva A M , Serrano A and Bortoli D 2015 Aerosol radiative effects during two desert dust events in August 2012 over the Southwestern Iberian Peninsula. *Atmos. Res.* **153**, 404–415. http://dx.doi.org/10.1016/j.atmosres.2014.10.007

[12] Gueymard C 2004 The sun's total and spectral irradiance for solar energy applications and solar radiation models. *Sol. Energy 76*, 423–453

[13] Stamnes K, Tsay S C, Wiscombe W and Laszlo I 2000 DISORT, a general purpose Fortran program for discrete-ordinate-method radiative transfer in scattering and emitting layered media: documentation of methodology. Goddard Space Flight Center, NASA

IOP Conf. Series: Earth and Environmental Science **28** (2015) 012011 doi:10.1088/1755-1315/28/1/012011

Trends in the aerosol load properties over south eastern Italy

J A G Orza[1] and M R Perrone[2]

[1]SCOLAb, Física Aplicada, Universidad Miguel Hernández, 03202 Elche, Spain.
[2]Department of Mathematics and Physics, University of Salento, 73100 Lecce, Italy.

E-mail: ja.garcia@umh.es

Abstract. The long-term (2003-2013) variations in columnar aerosol properties at Lecce, a site representative of the central Mediterranean, have been analysed for trend assessment. The study focuses on aerosol optical thickness (AOT) at 340, 440, 500 and 1020 nm and Ångström exponent (AE) for the pair 440-870 nm, retrieved from a sun photometer operating within the Aerosol Robotic Network (AERONET). A non-parametric trend analysis of the monthly mean, median and upper and lower tails (90th and 10th percentiles) suggests that the aerosol load has decreased during the study period, while the mean particle size remained unchanged. The characteristic advections reaching the study site were found by clustering analysis of back trajectories at 500, 1500 and 3000 m. Despite the strong influence they have on aerosol load and particle size, neither of the trends in advection routes could explain the tendencies found in the columnar aerosol properties. However, trends in aerosol data by advection type allow understanding the overall trends. Aerosol properties under flows with high residence time over continental Europe present differences according to the specific residing area. More specifically, no trend is found when flows arrive from Ukraine and the Balkans, while under advections from north-western/central Europe there are downward trends in the background levels and a reduction of the fine fraction. Negative trends are also found under flows with high residence time over the Mediterranean and northern Africa, again with differences according to the residing area.

1. Introduction

Atmospheric aerosols are an important component of the atmosphere due to their role in the Earth's radiation budget and in cloud and precipitation processes, and due also to their impact on air quality. Aerosol effects largely depend on the aerosol optical and microphysical properties. The long term analysis of such properties is of particular importance to investigate the changes of natural and anthropogenic aerosol contributions and to better evaluate the aerosol impact on the surface radiation balance.

Numerous studies on the temporal variation of aerosols and the assessment of trends have been conducted at different spatial and time coverage. Works using measurements at the global scale with satellite sensors and those dealing with data from a large number of monitoring stations may give (within intrinsic limitations and differences) a broad picture of the existence of trends in aerosol load and properties. To this end, studies of AOT measured by satellite sensors such as the Advanced Very High Resolution Radiometer (AVHRR) [1], the Sea-viewing Wide Field-of-view Sensor (SeaWiFS) [2], the Moderate Resolution Imaging Spectroradiometer (MODIS) [3, 4, 5] and the Multiangle Imaging Spectroradiometer (MISR) [4, 5] were performed, in addition to AOT studies using ground-based measurements by sun photometers at a large number of sites belonging to AERONET [6, 7], and aerosol optical parameters at the ground level retrieved with nephelometers and aethalometers within the GAW and the Interagency Monitoring of Protected Visual Environments (IMPROVE) networks [8]. Studies of long-term observations of PM10 concentrations at the EMEP (European Monitoring and Evaluation Programme) monitoring sites [9], and of aerosol number concentrations at GAW (Global Atmosphere Watch) and ACTRIS (Aerosols, Clouds, and Trace gases Research InfraStructure Network) sites [10] were also performed. These studies found, in general, downward trends over Europe and North America (mostly in their western and eastern parts, respectively), and mostly upward trends over South and East Asia. These tendencies are consistent with the emission-inventory assessments in these regions. As for the trends observed in dust emissions and export from source areas, it is found a positive trend in the Arabian Peninsula while in northern Africa the trend is negative.

We present an analysis of AOT and AE data for the period 2003-2013 at Lecce, south eastern Italy, for long-term trend assessment. The monitoring site can be considered as representative of coastal sites in the Central Mediterranean away from large sources of local pollution. The dependence of the columnar aerosol optical properties on the synoptic-scale advection patterns is analyzed. The observed strong dependence motivates the subsequent exploration for the existence of trends both in the frequencies of the major advection types and in the aerosol load and its optical properties by advection type.

2. Data and methods

A ten-year (March 2003-February2013) dataset of AOT at wavelengths 340, 440, 500 and 1020 nm and AE for the 440-870 nm pair is considered. Level 2 data, cloud screened and quality assured, were retrieved from direct solar measurements with a CIMEL CE 318 sun/sky photometer operating within AERONET at the University of Salento (40.33 N, 18.10 E, 27 m asl; Lecce, southeastern Italy) and are available at http://aeronet.gsfc.nasa.gov. The methods used to measure solar radiation and the instrument description are given, e.g. in [11]. The algorithms for cloud-screening and the retrieval of aerosol properties are described in [12] and [13]. The measurement error of the aerosol optical thickness is estimated to be in the range of 0.01–0.02 for the visible channels and 0.02 for UV [11, 14].

The assessment of monotonic trends was performed by non-parametric procedures based on the Mann-Kendall test in combination with the Theil-Sen (TS) slope estimate, over the time series of monthly mean, median, 10th and 90th percentile values. The Mann-Kendall test allows working with non-normal data and in situations with many missing values, and it is resistant to extreme cases. It, however, requires independence between observations; therefore, short-term persistence and seasonality alter the significance levels and may lead to wrong conclusions about the presence of a trend. The seasonal Kendall (seasK) test [15], as well as the trend-free pre-whitening (TFPW) procedure [16] applied over the seasonally adjusted monthly series (deseas+TFPW), were used. The first test is robust against seasonality but not against serial dependence, while the second one removes the lag-1 autocorrelation from the deseasonalized time series before applying the Mann-Kendall test. Two-tailed tests at 95 and 90% significance were conducted. The deseasonalized time series used in the TFPW procedure were obtained from a seasonal-trend decomposition of the series based on LOESS (locally weighted low-degree polynomial regression) applied recursively to the seasonal and trend components [17].

Kinematic 96-hour back-trajectories were calculated with the HYbrid Single-Particle Lagrangian Integrated Trajectory (HYSPLIT) model v 4.8 at 500, 1500 and 3000 m asl, every 6 hours (00, 06, 12, 18 UTC) during the study period. ERA-Interim data of 1.5° resolution was used as meteorological input. The trajectories were subsequently classified into homogeneous groups by a robust clustering procedure based on the k-means algorithm. Hourly longitudes and latitudes were the input variables and the similarity metrics was based on great-circle distances. A detailed description of the procedure can be found in [18] or [19]. Each trajectory was associated to the closest AERONET data in time to study aerosol properties by advection type.

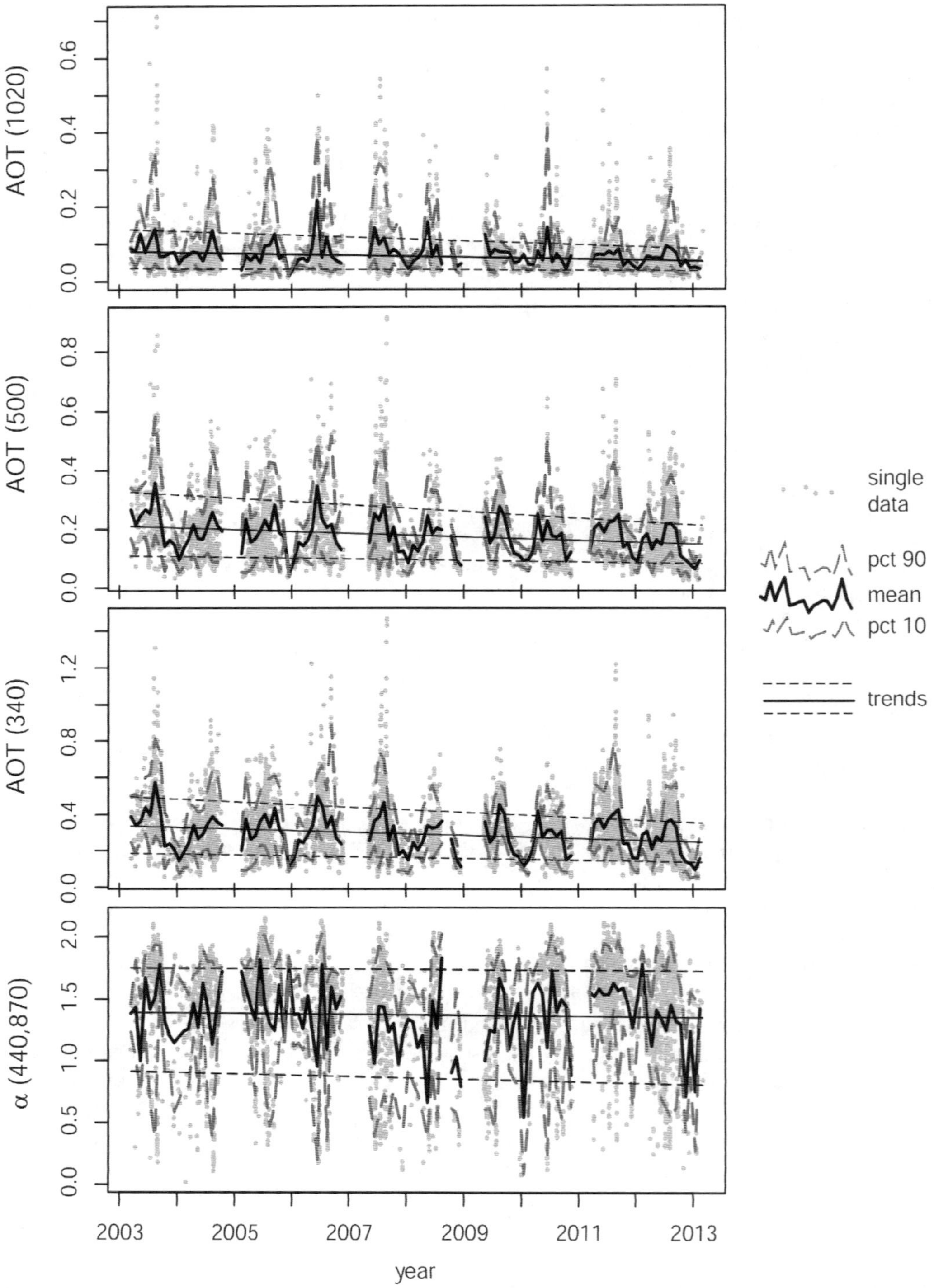

Figure 1. Single data (grey circles), monthly means and 10th and 90th percentile (pct10 and pct90, respectively) time series of AOT and AE in the study period. The linear trends were estimated with the Theil-Sen slope.

3. Results

3.1. AOT and AE trends

According to both tests there exist significant downward trends in all the AOT time series, at least at the 90% confidence level, while no trend is found in the AE calculated at the 440 and 870 nm wavelength pair (see Figure 1 and Table 1). These results suggest that the aerosol load has decreased over the study period while the mean particle size likely remained unchanged from 2003 to 2013.

The trends reported by Li et al. [7] for the AOT(440) and AE monthly medians are similar to the ones reported here. It is also observed (Table 1) that the (negative) AOT trends are stronger at lower wavelengths. In addition, at each wavelength the strongest trends are found for the 90th percentile time series while the weakest ones are found for the 10th percentile values. This means that the number of intense episodes have decreased over the study period, while the background aerosol load shows a much lesser lowering if any. The trends in the 90th percentile of the AOT at 1020 nm (that is more sensitive to coarse mode particles) and of the AOT at 340 nm (more sensitive to fine mode particles) would indicate that the concentration of both coarse and fine mode particles are decreasing over time.

Table 1. TS trend estimates for the monthly AERONET time series that present significant trends at 95% (in bold) or 90% confidence level according to the SeasK and deseas+TFPW tests. Trends were calculated on the original and the seasonally-adjusted series, respectively.

	seasK	deseas+TFPW
Aerosol parameter	Trend (decade^{-1})	Trend (decade^{-1})
Mean AOT(1020)	**-0.026**	**-0.023**
Median AOT(1020)	-0.017	**-0.013**
pct10 AOT(1020)	-0.008	**-0.007**
pct90 AOT(1020)	**-0.061**	**-0.065**
Mean AOT(500)	**-0.057**	**-0.051**
Median AOT(500)	**-0.052**	**-0.041**
pct10 AOT(500)		**-0.022**
pct90 AOT(500)	**-0.115**	**-0.010**
Mean AOT(440)	**-0.062**	**-0.053**
Median AOT(440)	-0.058	**-0.045**
pct10 AOT(440)		**-0.025**
pct90 AOT(440)	**-0.115**	**-0.103**
Mean AOT(340)	**-0.084**	**-0.073**
Median AOT(340)	**-0.084**	**-0.067**
pct10 AOT(340)	-0.050	**-0.040**
pct90 AOT(340)	**-0.133**	**-0.123**

3.2. Major advection patterns and trends in their monthly occurrence

Advection at the synoptic scale has a strong impact on the aerosol optical properties and mass concentration, as it has been shown at the study site for different, shorter, time periods [19, 20, 21]. The major advection routes were identified by a cluster analysis of back-trajectories, which resulted in 7, 6, and 7 major routes at 500, 1500 and 3000 m asl, respectively (see Figure 2). They are named according to their geographical pathway.

The airflows with a significant residence time over the Mediterranean are W_{500}/$WSW_{1500,3000}$, which cross the western Mediterranean after passing over the Iberian Peninsula or the northern-most part of Africa; and S which are slow-moving flows arriving from the southern side of the central Mediterranean basin. While WSW flows are common throughout the year with slightly higher frequency in spring and autumn, S are less frequent in the central months.

The advection routes with continental Europe characteristics are NE flows, which pass over Ukraine and the Balkans before reaching the study site and are more frequent in spring and autumn, while in March they are almost absent; and sNW_{500}/WNW_{1500}/$sWNW_{3000}$ slowly-moving flows that pass over the Italian Peninsula, are the most common type of advection and present strong seasonality due to their strong predominance in summer at the study site; and NW_{500} which cross over northwestern/central Europe after arriving from the North Sea.

Table 2. TS trend estimates for the monthly frequencies of occurrence of the advection patterns that present significant trends at 95% (in bold) or 90% confidence level according to the SeasK and deseas+TFPW tests. Trends were calculated on the original and the seasonally-adjusted series, respectively.

Parameter	seasK Trend (# decade^{-1})	deseas+TFPW Trend (# decade^{-1})
Freq S_{500}		-4.2
Freq NNW_{1500}	-3.2	-3.8
Freq NNW_{3000}	**-7.1**	**-7.7**
Freq $fWNW_{3000}$	2.0	
Freq $sWNW_{3000}$	4.6	5.8

There are two types of fast advections influencing SE Italy: NNW flows, which are of Arctic/continental polar nature, and WNW_{500}/$fWNW_{1500,3000}$ flows that correspond to polar maritime air masses traced back to north America. Both are the less frequent advection routes and show a strong seasonal behavior as are mostly found from November to March. NNW are more frequent in March, while the polar maritime air masses show a maximum in December/January.

Further details on the synoptic and the local meteorology associated to each advection route can be found in [20], with minor dissimilarities due to differences in the study time periods. Figure 3 shows the basic statistics of the AERONET aerosol properties by advection route. Three contrasting behaviors can be appreciated: (1) S and W_{500}/$WSW_{1500,3000}$ flows have the lowest AE and the highest AOT values, indicating loaded situations with coarse particles. (2) NE and sNW_{500}/WNW_{1500}/$sWNW_{3000}$ flows present the highest AE values and high AOT levels (these latter are higher the lower the wavelength and AOT340 are comparable to those of S and W_{500}/$WSW_{1500,3000}$) and correspond to fine particles of anthropogenic origin. (3) NNW, as well as WNW_{500}/$fWNW_{1500,3000}$ advections, show the lowest AOT and intermediate AE levels that are lower for the polar maritime air masses, and therefore are clean air masses.

The trend analysis of the monthly frequencies reveals monotonic trends in only a few advection types: NNW flows reaching the study site at 3000 m show downward trend (at the 95% confidence level) and also at 1500m (at the 90% confidence) with both analyses (Table 2). The relative reduction of NNW3000 is strong: 7 cases per decade for a mean of 11 cases per year. In turn, sWNW and fWNW at 3000 m present upward trends but at lower significance, as it is found for S flows at 500 m. Neither of these trends in advection routes explains the trends found in the columnar aerosol parameters.

IOP Conf. Series: Earth and Environmental Science **28** (2015) 012011 doi:10.1088/1755-1315/28/1/012011

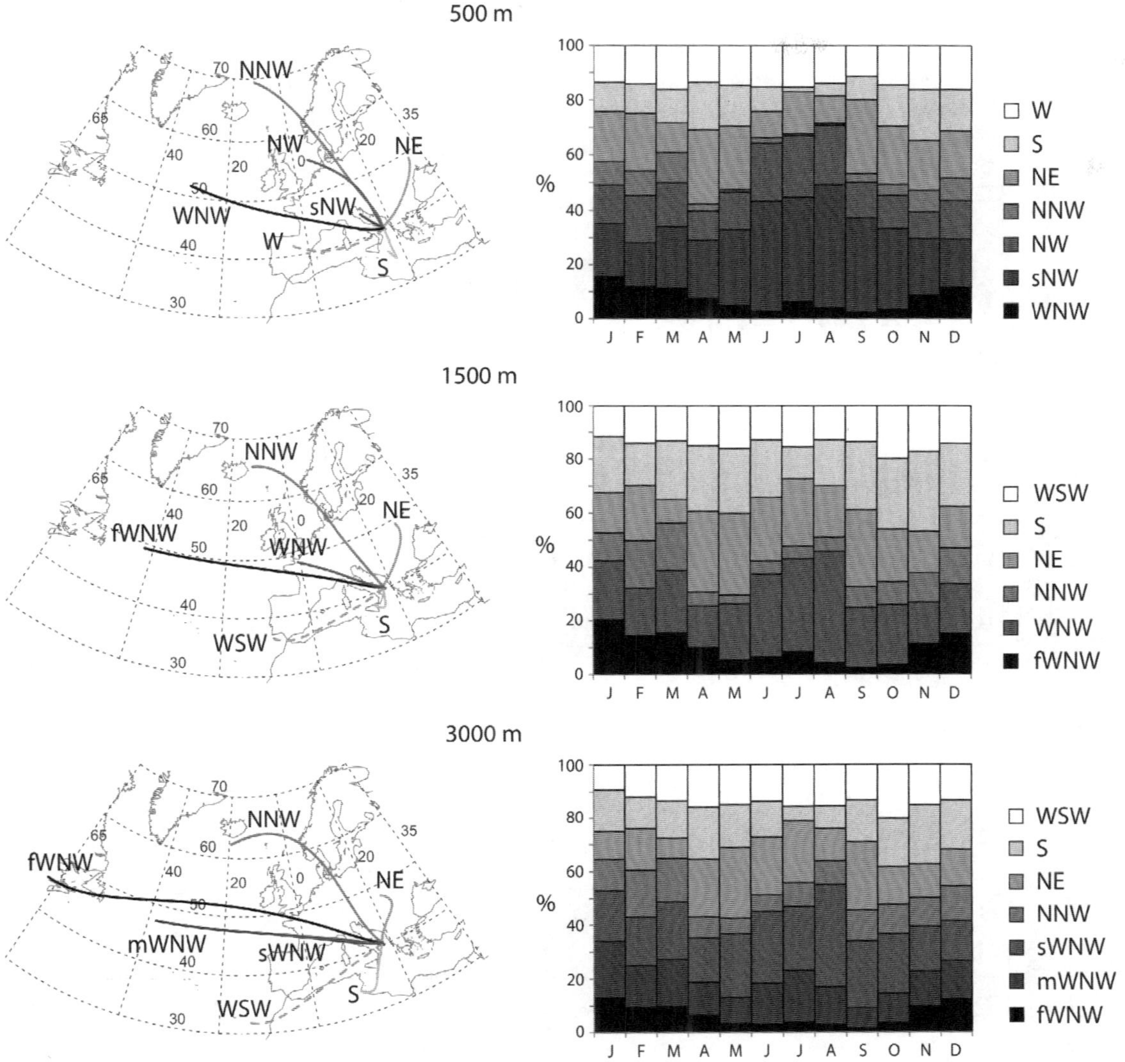

Figure 2. Centroids of the major advection patterns reaching the study site at 500, 1500 and 3000 m (left) and their respective monthly frequencies of occurrence (right).

3.3. Trends in AOT and AE by advection route

We focus on the results for the advections reaching the study site at 500 and 3000m. It should be noted that in some cases the SeasK test could not be applied to the data segregated by advection type due to the lack of data in some seasons; the analysis in these cases is made with the deseas+TFPW procedure and it is checked that the confidence intervals of the TS slope do not include the zero slope.

The analysis of the aerosol parameters by advection at 500 m reveals (Table 3) mostly negative trends both under advections associated to fine aerosols and under advections associated to the coarse fraction. AOT values under sNW and NW flows present downward trends; however, no trend is found under NE flows which are also associated to anthropogenic aerosols. Under sNW, trends in mean and median AOT are similar to those of the overall AOT at 1020 and 500 nm, although they become lower and less significant at lower wavelengths. The trends in the AOT upper tails (90th percentile) may indicate a reduction of episodic events during sNW advections. Under NW flows, the 10th percentile AE is found to have a significant negative trend (95% confidence) by the two tests; similarly, the 10th percentile of all the AOT values show a decreasing trend at the 95% confidence after the deseas+TFPW test, with trend magnitudes that decrease with decreasing wavelength. This may indicate a reduction in the background concentrations and also a reduction of the fine mode fraction. In turn, the lower tails in NNW show positive trends of small magnitude after the deseas+TFPW test.

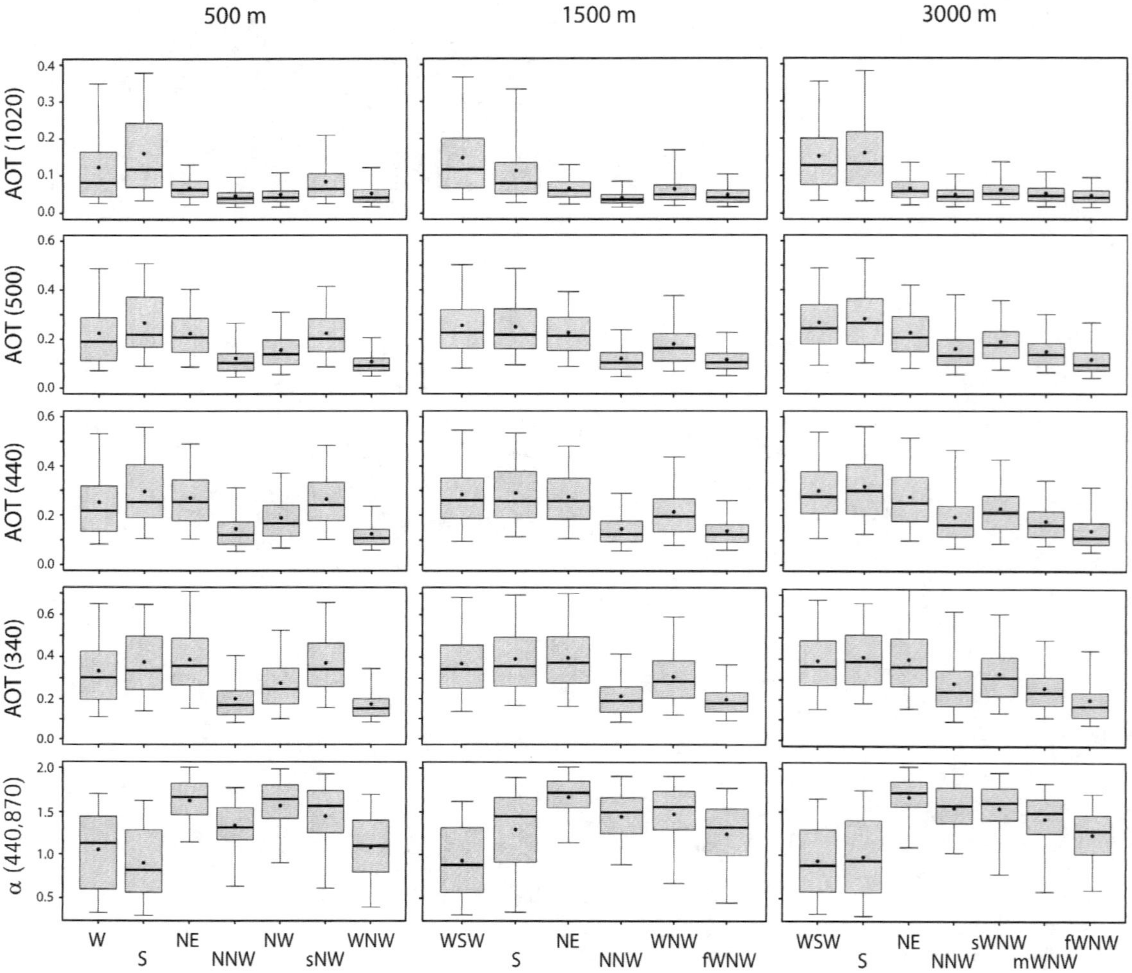

Figure 3. AOT and AE levels by advection type. Mean (dot), median (horizontal bold line), 25th and 75th percentiles (lower and upper box boundaries) and 5th and 95th percentiles (whiskers) are shown.

Regarding the aerosol properties under advections associated to coarse particulates like African dust and sea-salt particles, the mean, median and 10th percentile AOT at 500, 440 and 340 nm under S flows present negative trends of much higher magnitude than the overall AOT trends, while the trends for the upper tails are of the same magnitude. Under W flows, trends behave mostly like S advections but with higher significance for the AOT(1020) values.

Trends in the aerosol parameters by advections reaching Lecce at 3000m show some differences (Table 4): no trend is found under S advections, while under WSW there are negative trends in all the AOT parameters and an upward trend in the 90th percentile AE that would imply a reduction in dust load associated to these flows and also a decrease in the coarse-dominated episodes. The AE parameters present downward trends also under NNW (mean and 90th percentile) and fWNW advections (all the parameters). While NNW show no trends in AOT, fWNW present trends in AOT(340) and AOT(440) and mWNW flows present downward trends in all the AOT parameters.

Finally, like in the case of S flows, the other two types of slower advections, i.e. NE and sWNW, show no trend.

Table 3. TS trend estimates for the monthly time series by advection type reaching the study site at 500 m, with significant trends at 95% (in bold) or 90% confidence.

	seasK	deseas+YP		seasK	deseas+YP
Aerosol parameter	Trend (decade^{-1})		Aerosol parameter	Trend (decade^{-1})	
W_{500}			**NNW_{500}**		
mean AOT(1020)	**-0.052**	**-0.077**	pct10 AOT(1020)		**0.009**
median AOT(1020)	**-0.049**	**-0.076**	pct10 AOT(500)		**0.018**
pct10 AOT(1020)	**-0.040**	**-0.052**	pct10 AOT(440)		**0.026**
pct90 AOT(1020)	**-0.060**	-0.112	pct10 AOT(340)		**0.032**
mean AOT(500)	**-0.105**	**-0.140**			
median AOT(500)	**-0.093**	-0.128	**NW_{500}**		
pct10 AOT(500)	**-0.072**	-0.088	pct10 AOT(1020)		**-0.010**
pct90 AOT(500)	**-0.115**		pct10 AOT(500)		**-0.037**
mean AOT(440)	**-0.113**	-0.146	pct90 AOT(500)		-0.050
median AOT(440)	**-0.104**	-0.141	pct10 AOT(440)		**-0.042**
pct10 AOT(440)	**-0.082**	**-0.101**	pct90 AOT(440)		-0.052
pct90 AOT(440)	**-0.127**		pct10 AOT(340)		**-0.063**
mean AOT(340)	**-0.138**		Mean AE(440,870)		-0.184
median AOT(340)	**-0.129**	-0.177	pct10 AE(440,870)	**-0.418**	**-0.355**
pct10 AOT(340)	**-0.118**	**-0.135**			
pct90 AOT(340)	**-0.150**		**sNW_{500}**		
			mean AOT(1020)	**-0.026**	**-0.022**
S_{500}			median AOT(1020)	**-0.026**	**-0.021**
pct10 AOT(1020)		-0.040	pct90 AOT(1020)	**-0.046**	**-0.040**
mean AOT(500)		**-0.108**	mean AOT(500)	**-0.060**	**-0.048**
median AOT(500)		**-0.115**	median AOT(500)	**-0.055**	-0.047
pct10 AOT(500)		**-0.102**	pct10 AOT(500)	-0.022	
mean AOT(440)		**-0.106**	pct90 AOT(500)	**-0.087**	**-0.073**
median AOT(440)		**-0.124**	mean AOT(440)	**-0.061**	**-0.050**
pct10 AOT(440)		**-0.107**	pct90 AOT(440)	**-0.094**	**-0.079**
pct90 AOT(440)		-0.112	mean AOT(340)	-0.074	-0.068
mean AOT(340)		**-0.140**	median AOT(340)	-0.073	
median AOT(340)		**-0.157**	pct10 AOT(340)	**-0.042**	
pct10 AOT(340)		**-0.135**	pct90 AOT(340)	-0.112	**-0.102**
pct90 AOT(340)		**-0.133**			
			WNW_{500}		
			pct90 AOT(340)		**-0.067**

4. Summary and conclusions

The long-term variations in aerosol properties at a site representative of the central Mediterranean have been explored in this work. A nonparametric trend analysis of monthly statistics of AOT and AE gives statistical evidence that the aerosol load has decreased over 2003-2013 in SE Italy, while the mean particle size remained unchanged.

The synoptic-scale advections have a strong influence on the aerosol load and particle size. However, their trends in frequency of occurrence cannot explain the observed trends in the aerosol properties. It should be noted that meteorology shows a strong interanual variability that is not well described in terms of linear tendencies. The trends in AERONET aerosol properties by advection type allow to understand the origin of the trends in the overall time series. Trends under advections with high residence time over continental Europe present clear differences depending on their specific residing area: no trend is found with flows from Ukraine and the Balkans (NE), while aerosols under advections that arrive from anywhere in the Italian Peninsula (sNW) present trends similar to the overall AOT series, and advections from north-western/central Europe (NW) show downward trends in the background levels and a reduction of the fine fraction that would be a signal of the reduction of the emissions in EU. The flows with high residence time over the Mediterranean, which may also carry African dust to the study site, present also some differences in the corresponding aerosol trends: the aerosol load have decreased under W_{500}/WSW_{3000} advections while such a reduction is found only under S at the lowest level. Fast advections, of polar continental/arctic or polar maritime origin show a reduction in the fine fraction when they arrive at 3000 m; and while the WNW_{500}/$fWNW_{3000}$ show a decrease in AOT(340), there is an increase in the background aerosol load under NNW reaching the site at the lowest height.

The present exploratory study highlights that the analysis of the aerosol properties at a site by large-scale advection routes may show separately the long term changes in emissions at different source areas.

Future work on PM10 mass concentration may provide insight on the presence of aerosols inside/above the ABL and would help to further understand the causes of long-term variation. However, differences in sampling scheme with respect to the columnar measurements, as well as the influence of cloud disturbance and onset and end times of pollution episodes on the latter need to be accounted for to properly compare measurements.

Table 4. TS trend estimates for the monthly time series by advection type reaching the study site at 3000 m, with significant trends at 95% (in bold) or 90% confidence.

Aerosol parameter	deseas+TFPW Trend (decade^{-1})	Aerosol parameter	seasK Trend (decade^{-1})	deseas+TFPW Trend (decade^{-1})
WSW_{3000}		**$mWNW_{3000}$**		
mean AOT(1020)	**-0.063**	mean AOT(1020)	**-0.030**	**-0.027**
median AOT(1020)	**-0.075**	median AOT(1020)	**-0.024**	**-0.021**
pct10 (AOT1020)	**-0.054**	pct10 AOT(1020)	**-0.024**	**-0.026**
pct90 AOT(1020)	**-0.063**	pct90 AOT(1020)	**-0.042**	**-0.033**
mean AOT(500)	**-0.111**	mean AOT(500)	**-0.083**	**-0.072**
median AOT(500)	**-0.116**	median AOT(500)	-0.068	**-0.056**
pct10 AOT(500)	**-0.086**	pct10 AOT(500)	-0.056	**-0.057**
pct90 AOT(500)	**-0.114**	pct90 AOT(500)	**-0.116**	**-0.096**
mean AOT(440)	**-0.115**	mean AOT(440)	**-0.096**	**-0.081**
median AOT(440)	**-0.123**	median AOT(440)	-0.079	**-0.066**
pct10 AOT(440)	**-0.093**	pct10 AOT(440)	-0.065	**-0.067**
pct90 AOT(440)	**-0.120**	pct90 AOT(440)	**-0.132**	**-0.109**
mean AOT(340)	**-0.140**	mean AOT(340)	**-0.139**	**-0.119**
median AOT(340)	**-0.150**	median AOT(340)	**-0.122**	**-0.106**
pct10 AOT(340)	**-0.118**	pct10 AOT(340)	**-0.106**	**-0.109**
pct90 AOT(340)	**-0.162**	pct90 AOT(340)	**-0.183**	**-0.161**
pct90 AE(440,870)	**0.181**			
		$fWNW_{3000}$		
NNW_{3000}		median AOT(440)		**-0.056**
mean AE(440,870)	**-0.194**	mean AOT(340)		**-0.071**
median AE(440,870)	-0.213	median AOT(340)		**-0.086**
pct90 AE(440,870)	**-0.194**	pct10 AOT(340)		**-0.054**
		mean AE(440,870)		**-0.287**
		median AE(440,870)		**-0.270**
		pct10 AE(440,870)		**-0.276**
		pct90 AE(440,870)		**-0.268**

Acknowledgement
The authors gratefully acknowledge the NOAA Air Resources Laboratory (ARL) for the provision of the HYSPLIT transport and dispersion model, and A. Stein for providing the ERA-Interim data used as input for the trajectory calculations.

References
[1] Mishchenko M I, Geogdzhayev I V, Liu L, Ogren J A, Lacis A A, Rossow W B, Hovenier J W, Volten H and Muñoz O 2003 *J. Quant. Spectrosc. Radiat. Transfer* **79-80** 953-72, doi:10.1016/S0022-4073(02)00331-X
[2] Hsu N C, Gautam R, Sayer A M, Bettenhausen C, Li C, Jeong M J, Tsay S-C, and Holben B N 2012 *Atmos. Chem. Phys.* **12** 8037-53, doi:10.5194/acp-12-8037-2012
[3] Kishcha P, Starobinets B and Alpert P 2007 *Geophys. Res. Lett.* **34** L05810
[4] Meij A, Pozzer A and Lelieveld J 2010 *Atmos. Chem. Phys. Discuss.* **10** 30731–76
[5] Zhang J and Reid J S 2010 *Atmos. Chem. Phys.* **10** 10949-63, doi:10.5194/acp-10-10949-2010
[6] Xia X A 2011 *Environ. Res. Lett.* 2011 **6** 044011
[7] Li J, Carlson B E, Dubovik O and Lacis A A 2014 *Atmos. Chem. Phys.* **14** 12271-89, doi:10.5194/acp-14-12271-2014
[8] Collaud Coen M et al. 2013 *Atmos. Chem. Phys.* **13** 869-94, doi:10.5194/acp-13-869-2013
[9] Tørseth K, Aas W, Breivik K, Fjæraa A M, Fiebig M, Hjellbrekke A G, Lund Myhre C, Solberg S and Yttri K E 2012 *Atmos. Chem. Phys.* **12** 5447-81, doi:10.5194/acp-12-5447-2012
[10] Asmi A et al. 2013 *Atmos. Chem. Phys.* **13** 895-916, doi:10.5194/acp-13-895-2013
[11] Holben B N et al. 1998 *Remote Sens. Environ.* **66** 1–16
[12] Dubovik O and King M D 2000 *J. Geophys. Res.* 105 D**16** 20673–96
[13] Smirnov A, Holben B N, Eck T F, Dubovik O and Slutsker I 2000 *Remote Sens. Environ.* **73** 337–49
[14] Eck T F, Holben B N, Reid J S, Dubovik O, Smirnov A, O'Neill N T, Slutsker I and Kinne S 1999 *J. Geophys. Res.* **104** 31333–50
[15] Hirsch R M et al. 1982 *Water Resour. Res.* **18** 107–21
[16] Yue S et al. 2002 *Hydrol. Process.* **16** 1807–89
[17] Cleveland R B et al. 1990 *J. Off. Stat.* **6** 3-73
[18] Orza J A G, Cabello M, Galiano V, Vermeulen A T and Stein A F 2012 *Lagrangian Modeling of the Atmosphere* eds J Lin et al. (Washington: American Geophysical Union) p 127-41, doi: 10.1029/2012GM001315
[19] Perrone M R, Becagli S, Garcia Orza J A, Vecchi R, Dinoi A, Udisti R and Cabello M 2013 *Atmos. Environ.* **71** 176-186
[20] Perrone M R, Romano S and Orza J A G 2014 *Atmos. Res.* **145-146** 152-167
[21] Perrone M R, Romano S and Orza J A 2015 *Environ. Sci. Pollut. Res.* doi: 10.1007/s11356-015-4850-7

IOP Conf. Series: Earth and Environmental Science **28** (2015) 012012 doi:10.1088/1755-1315/28/1/012012

Review of critical parameters in biomass combustion emissions control by means of hybrid filter

D Sanz[1], E Rojas[1], J J Rodríguez-Maroto[1], R Ramos[2], E Borjabad[2], R Escalada[2], S García-Alonso[1], C Gutierrez-Canas[3], G Aragon[3], I Mugica[3], I Ibarra[3], I Celades[4] and V Sanfelix[4]

[1] CIEMAT Ave Complutense 40, 28040 Madrid, Spain
[2] CEDER-CIEMAT Motorway A-15 exit 56, 42290 Lubia (Soria), Spain
[3] University of the Basque Country UPV-EHU Alameda Urquijo s/n, 48013 Bilbao, Spain
[4] Ceramic Technology Institute Ave Vicent Sos Baynat s/n, 12006 Castellon, Spain

E-mail: david.sanz@ciemat.es

Abstract. Control of particulate matter emissions by means of hybrid filter has been included in the experimental plan of a research project in the field of Mediterranean agro-forestry waste biomass combustion at medium scale. Application of hybrid filters to biomass combustion has not been thoroughly experimented so far. An identification of the most important parameters in particulate matter emissions control by means of a hybrid filter was undertaken. The filter involves two of the most significant technologies in fly ash emission control, electrostatic precipitation and fabric filtration A discussion of these parameters and principles of operation of said technologies is presented, as well as the final selection of parameters to be included in the experimental matrix of the project in regards to emissions control. A novel approach is proposed for testing of filtration velocity influence in fabric filter module without impacting on operation of electrostatic precipitation module.

1. Introduction

The project "Integrated strategy to predict, monitor and ensure the sustainability of the combustion of agricultural and forestry waste biomass"(CLEANBIOM) is intended to facilitate the management and sustainable operation of medium scale decentralized combustion units, firing Mediterranean agro-forestry waste biomass. Its general objective is to develop a comprehensive strategy, for the prediction, control, and minimization of the pollutants generated in the combustion process of waste biomass in the Mediterranean basin region. Fly-ash is prominently included among these pollutants. Mediterranean biomass may show significant differences in composition from North European sources, since species are different. Also, biomass waste consists mostly on small branches and bark witch show composition peculiarities compared to other plant parts [1].

The experimental data on combustion and emissions control will be obtained in CEDER-CIEMAT, on a scale comparable to actual production plants, aimed to combine comprehensive and size segregated characterization of primary and secondary atmospheric pollutants, ashes and precursors of corrosion. The pilot plant incorporates a bubbling fluidized bed combustion plant (1 MW nominal output) and emissions treatment line provided with control sensors and sampling stations. This line has a hybrid filter that integrates electrostatic precipitation (dry, wire-parallel plates) and fabric filter modules, which has been previously used in other research projects [2]. Very little research work has been done so far on the application of hybrid filters to biomass combustion [3].

The project work plan includes an exhaustive identification of all and each one of the parameters that can influence the process of combustion (efficiency) and emissions and their control through the application of different technologies. Afterwards, a plan of experiments that will ensure the attainment of objectives will be designed. In this sense, the critical parameters in the control of particulate matter emissions have been identified. A necessarily limited selection of the most relevant parameters for inclusion in the experimental matrix of the project has been conducted.

2. Fundamentals of electrostatic precipitators and fabric filters operation

2.1. Electrostatic precipitators

The operation of an electrostatic precipitator (ESP) involves basically two physical phenomena [4]:

- Electric charging of particles (by diffusion and field charging).
- Movement (migration) of charged particles in an electric field to the collection electrodes.

The charging mechanism is based on the migration of ions present in the gas towards the surface of the particles, either by Brownian diffusion (key for particles smaller than 0.2 μm), or driven by electric field (key for particles larger than 1 μm). Ions are produced in the precipitator continuously by a corona discharge (localized electrical breakdown) in the region near the emitting electrodes.

The simplest approach at ESP efficiency modeling is based on a material balance on particles in a cross section of the precipitator. The following expression for overall collection efficiency of an electrostatic precipitator, known as the Deutsch-Anderson equation [5], is derived:

$$\eta = 1 - \exp\left[-w\left(\frac{A}{Q}\right)\right]$$

where:

η : Mass collection efficiency of the precipitator (ratio between the mass concentrations of fly ash at the exit and at the entrance of the precipitator $\eta = C_1/C_0$).

w : Migration velocity of particles (terminal velocity of charged particles in the electric field)

A : Surface area of collection electrodes exposed to electric field.

Q : Volume flow rate of gas through the precipitator.

Migration velocity, from theoretical considerations, should be proportional to the diameter of the particle and to the product of charging and collecting electric fields (in single-stage precipitators one single field plays both roles), as shown in the following expressions, applicable to coarse and fine particles, respectively:

$$w = \frac{\varepsilon\varepsilon_0}{\varepsilon + 2}\frac{d_p E_c E_m}{\mu}$$

$$w = \varepsilon_0 \frac{d_p E_c E_m}{\mu} C_C \left[\left(1 + \frac{2\lambda}{d_p}\right)^2 + \frac{2}{1 + \frac{2\lambda}{d_p}} \frac{\varepsilon - 1}{\varepsilon + 2}\right]$$

where:

ε, ε_0 : Particle and vacuum dielectric constant, respectively

μ : Gas viscosity
λ: Gas mean free path
d_p : Particle diameter
C_C: Cunningham correction factor
E_c, E_m : Charging and collecting electric field, respectively.

In industrial practice, use of Deutsch-Anderson model presents constraints due to its numerous assumptions (plug flow, absence of particle diffusion, uniform distribution of particles) and the lack of consideration for practical adverse phenomena (bypass flow, collected ash resuspension, etc...). In addition, fly ash is usually polydisperse rather than of a uniform particle size. Migration velocities are difficult to estimate on a purely theoretical basis, thus experimental values obtained for aerosols of different nature are essential to be used.

A model similar to the Deutsch was proposed by Matts-Ohnfeldt [6]:

$$\eta = 1 - \exp\left[-w_k\left(\frac{A}{Q}\right)^k\right]$$

where:
w_k : Average migration velocity
k: Constant, usually between 0.4 and 0.6

This model modifies the relationship between the specific collection area (*A/Q*) and precipitator efficiency, reflecting the fact that, in practice, the area increases needed for a given improvement of efficiency are greater than those estimated by the Deutsch model.

Since the modern development of electronics provided widespread access to massive calculation capabilities, many procedures based on computational flow modelling have been used for the study and design of electrostatic precipitators,. However, in order to discuss the influence of the various factors involved, simple models such as those presented are enough.

2.2. Fabric filters

In fabric filter collectors dust is removed from the gas stream when dust-laden gas flows through a fabric. Woven fabrics or felts can be used, sometimes superficially coated by a porous membrane [7]. A wide range of materials is available (nylon, PTFE, etc...). A plate holding the sleeves or bags of filter media divides the filter module in raw and clean gas.

As filtration proceeds, retained dust forms a dust layer on the surface of filter media, known as filter cake, thus increasing the pressure drop across the filter. Therefore, it is necessary to periodically proceed to the cleaning of the filter, removing the dust cake to begin a new cycle of filtration.

According to the cleaning method, three sorts of fabric filter designs can be distinguished: shake, reverse flow and air pulse cleaning filters. In pulse cleaning filters the open ends of the sleeves are placed at the top. The gas flows from the outside towards the inside of the sleeves, leaving the dust retained on the sleeve's outer surface. A brief pulse of compressed air is released during the cleaning, typically in the order of 0.1 second, towards the inside of each sleeve through a nozzle located over its open end. In this way, a reverse flow of clean gas is induced through the fabric. Pulse air and clean gas expand the sleeve, forcing the release of the deposited dust. The pulse pressure is sufficient to dislodge dust without interrupting the flow of gas through the filter.

Particle penetration tends to be extremely low, except immediately after filter cleaning. There are models for the estimation of particle filtration efficiency in a porous medium [8]. However, in contrast to the electrostatic precipitator, filtration efficiency has no obvious relationship with the sizing of the equipment. The models usually used for filter sizing are those who estimate the pressure drop across the filter. The process involves the specification of gas flow rate and average pressure drop, from which the minimum required filter media area is determined. The most widespread model for the relationship between pressure drop and gas flow rate treated in a filter sleeve is the following [9]:

$$\Delta P = \Delta P_r + K.W.\frac{Q}{A}$$

where:
ΔP : Pressure drop
ΔP_r : Residual pressure drop of a newly cleaned sleeve. Dependent on filtration velocity, the nature of the filter media and cleaning conditions.
K : Specific resistance of the dust cake
W : Surface mass load of dust over the filter media. Dependent on the dust concentration in the raw gas, filtration velocity and the time elapsed since the last cleaning.

The Q/A ratio is usually known as gas-to-cloth ratio or filtration velocity.

3. Critical parameters associated with the operation of electrostatic precipitators

As we have seen, the efficiency of an electrostatic precipitator depends on a number of basic parameters:

- Treated gas flow rate (gas velocity or residence time).
- Collection area. This parameter and the former are often grouped into what is called specific collection area (Q/A), the collection area to gas flow ratio.
- Applied electric potential.

Essentially, these are the parameters included in Deutsch, Deutsch-Anderson and Matts-Onhfeld equations for estimation of efficiency.

In addition to influencing the efficiency through specific collection area, it has been found that the *axial velocity of the gas* between the electrodes has an influence on migration velocity [10]. The typical minimum gas velocity is 1 m/s. Migration velocity increases with increasing axial gas velocity up to ≈2 m/s. For higher gas velocities resuspension of collected ash results in lower effective migration velocity.

Terminal migration velocity is conditioned by the *physico-chemical properties* of gas and particles, which determine their dielectric properties and the particle charging process. In this sense, and in addition to the variables listed in the equations, there is another fundamental parameter for the operation of the precipitators, the *resistivity of the particles*. This has two basic components: volumetric conductivity and surface conductivity. The first depends on the chemical composition of the particles and the second on the species adsorbed on the surface and the gas composition (water vapor, sulfur oxides, ammonia, and some other compounds). Resistivity also depends on the temperature. Increasing temperature increases electrical resistivity up to a maximum. Then, total resistivity drops as a result of lower surface resistivity at higher temperatures. For a given type of particle emission, the usual way of controlling the resistivity is by means of operating temperature of the gas and, on occasions, through the modification of the composition of the gas. To that end, injection of agents such as ammonia, water vapor or SO_3 has been used.

When the resistivity increases above 10^{11} Ω cm, dust deposited on the collection electrode releases its electrical charge very slowly, which gives rise a potential difference through the layer of dust which partially neutralizes the electric field in the precipitator. The so-called reverse ionization or back corona occurs in extreme cases. Both effects adversely affect the efficiency of the precipitator. Low resistivities result in resuspension when collected ash easily releases its electrical charge, thus significantly decreasing the electrostatic force that holds ash to the collection electrode.

Regarding the resistivity values in the case of emissions from biomass combustion [11, 12], significant differences have been found between the soot, condensable organic compounds and inorganic matter. Condensable organic matter presents the highest resistivity (10^{10} Ωcm at 150°C). That of soot is the lowest (10^{6} Ωcm at 150°C). Inorganic matter resistivity values are intermediate, very close to the ideal for the operation of electrostatic precipitators (10^{9} Ωcm at 150°C). However, the resistivity of the inorganic matter at temperatures under 120°C is reduced below the ideal value, approaching the lower acceptable limit for the operation of electrostatic precipitators.

Regarding the influence of the *granulometry of ash particles* of, fine particles hinder collection and favor resuspension. Electrostatic precipitators typically have a minimum fractional efficiency in the range between 0.4 - 0.5 μm [13].

4. Critical parameters associated with the operation of fabric filters

Too high *filtration velocity* leads to higher average pressure drop, which increases penetration of particles through the filter (lower efficiency) and necessitating a more frequent cleaning, which negatively affects the life-span of the filter media. On the other hand, too low filtration velocity results in excessively large and expensive equipment. So, this parameter influences both investment and operation costs. It also alters the balance between depth and surface filtration mechanisms, influencing the penetration of particles through the filter.

Regarding *particle size distribution* of the ash, finer particles lead to the formation of more densely packed cakes resulting in higher pressure drop for a given filtration velocity. The nature of the particles can be problematic, if adhesive or capable of strong electrostatic interactions that attach them permanently to the filter media. The impossibility of removing them will prevent the continuation of the filtration process.

With regard to the *composition and temperature of the gas*, it is crucial to avoid the condensation of water vapor that would result in an increase of pressure drop, and possible corrosion problems. On the other hand, the operating temperature must be below the maximum tolerated by the material of the filter media.

5. Parameters selected for experimental trials

To select the parameters to be tested it must be taken into account which of them show the highest potential influence, as well as the practicality of manipulation in the experimental facility, and which parameters are already given or limited by the type of fuel used, the combustion conditions, etc. For example, since the hybrid filter processes the whole of the combustion plant flue gas, it is not easy to modify the treated gas flow rate. The total collection area of the electrostatic precipitator is not easy to modify. Two parameters have been finally selected for trials in the project: the potential applied in the electrostatic precipitator and the filtration velocity in the fabric filter.

The first parameter affects various aspects of the process, such as the relative distribution of the captured fly ash between two modules in the hybrid filter, or the energy consumption of the process. At the same time, the proportion of fly ash passing to the fabric filter module affects the dynamics of this module, its energy consumption, the penetration of particulate matter through the filter media, etc. This is because, through migration velocity, the applied potential crucially influences the efficiency of electrostatic precipitator [14].

As stated, filtration velocity in the fabric filter module is another fundamental parameter. The usual approach of varying gas flow rate is problematic. Thus, to vary the filtration velocity the available surface area of the filter will be reduced by removing some filter bags in some experiments. Moreover, this procedure is advantageous in that it has no impact on the electrostatic precipitator module operation.

Different fuels will be tested within the project. This presumably will give rise to different combustion gas compositions, particularly in what refers to water vapor, and perhaps to fly ash of different characteristics (particle size distribution, chemical composition, resistivity, etc...). So the occasion to observe the effects of other factors could present itself.

Acknowlegdements

CLEANBIOM project is partially funded by Ministry of Economy and Competitiveness of the Government of Spain within the framework of the scientific and technical research and innovation plan 2013-2016.

References

[1] Van Loo S and Koppejan J 2010 *The handbook of biomass combustion & co-firing* (London: Earthscan) chapter 2
[2] Aragon G et al 2015 Enhanced control of fine particle emissions from waste biomass combustion using hybrid filter *Energy & Fuels* **29** (4) 2358-2371
[3] Singh R and Shukla A 2014 A review on methods of flue gas cleaning from combustion biomass *Renewable and Sustainable Energy Reviews* **29** 854-864
[4] White H J 1977 Electrostatic precipitation of fly ash - Parts I, II, III and IV *Journal of the Air Pollution Control Association* **27** (1-4)
[5] Deutsch W 1922 Bewegung und Ladung der Elektrizitatstrager im Zylinderkondensator *Annals of Physics* **68** 335-344
[6] Matts S and Ohnfeldt P O 1964 *Efficient gas cleaning with SF precipitators* (Flakten)
[7] Cooper C D and Alley F C 2002 *Air Pollution Control* (Long Grove, Illinois: Waveland Press) chapter 6
[8] Hinds W C 1999 *Aerosol technology* (NewYork: John Wiley & Sons) chapter 9
[9] Dennis R and Klemm H A 1980 Modeling concepts for pulse jet filtration *JAPCA* **30** (1) 38-43
[10] Riehle C and Löffler F 1992 The effective migration rate in electrostatic precipitators *Aerosol Science and Technology* **16** (1) 1-14
[11] Nussbaumer T 2011 Characterisation of particles from wood combustion with respect to health relevance and electrostatic precipitation *Proc. of the 3rd Mid-European Biomass Conference (Graz)* (Austrian Biomass Association)
[12] Lillieblad L et al. 2004 Conditions for electrostatic precipitators after biomass fired boilers. *International Society for Electrostatic Precitators (ICESP) IX Conference*. May 17-21, 2004. Mpumalanga (South African Republic)
[13] Strand M et al 2002 Fly ash penetration through electrostatic precipitator and flue gas condenser in a 6 MW biomass fired boiler *Energy & Fuels* **16** 1499-1506
[14] Rodriguez-Maroto J J et al 1997 On the electrostatic precipitator resolution in function of main influence factors studied in a pilot fluidized bed coal combustion plant *Proc. ICCS'97 9th International Conference on Coal Science* September 7-12 Essen (Germany)

Multi-wavelength aerosol LIDAR signal pre-processing: practical considerations

A Rodríguez-Gómez[1], F Rocadenbosch[1], M Sicard[1], D Lange[2], R Barragán[1], O Batet[3], A Comerón[1], M A López Márquez[1], C Muñoz-Porcar[1], J Tiana[1] and S Tomás[4]

[1]Barcelona Lidar Station, Remote Sensing Laboratory (RSLAB), Department of Signal Theory and Communications, BarcelonaTech University (UPC), Spain.
[2]Departamento de Ciencias Exactas e Ingeniería, Universidad Católica Boliviana San Pablo, Bolivia
[3]Expert Ymaging S.L., IC3 - Catalan Institute of Climate Sciences
[4]Instituto de Ciencias del Espacio, CSIC, Spain.

E-mail: alejandro@tsc.upc.edu

Abstract. Elastic lidars provide range-resolved information about the aerosol content in the atmosphere. Nevertheless, a number of pre-processing techniques need to be used before performing the inversion of the detected signal: range-correction, time-averaging, photon-counting channel dead-time correction, overlap correction, Rayleigh-fitting and gluing of both channels.

1. Introduction

Elastic lidars provide range-resolved information about the aerosol content in the atmosphere [1]. Before performing the signal inversion, that will provide this information, a number of pre-processing techniques need to be performed. This paper provides a practical approach to some of them.

2. Range-correction and time-averaging

Figure 1 shows the system description of UPC 6-channel elastic lidar [2]. It includes a Nd:YAG laser transmitter with second and third harmonic generators, that produces pulses at 1064 nm, 532 nm and 355 nm. The receiving system includes a 36-cm telescope, a polychromator unit and six receiving channels: 1064, 607, 532, 407, 387 and 355 nm; in every channel both power (analogic) and photon counting reception is performed. For the rest of this paper we will be dealing with the received signal at 532 nm.

Figure 2 shows the received signal for a one-pulse transmitted signal. Different features must be pointed out: most of the signal is negligible, apart from a non-zero offset in the analog channel. According to the lidar equation [1]:

$$P(R)=\frac{E_{pulse}A_{teles}c}{2R^2}\left[\beta(R)\right]\cdot e^{-2\int_0^R\left[\alpha(x)\right]dx} \quad (1).$$

IOP Conf. Series: Earth and Environmental Science **28** (2015) 012013 doi:10.1088/1755-1315/28/1/012013

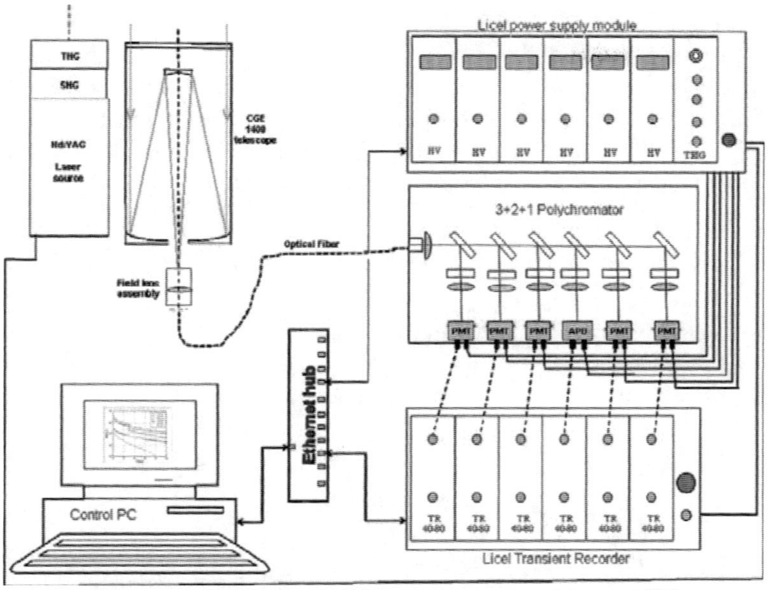

Figure 1. UPC elastic lidar system general layout [2].

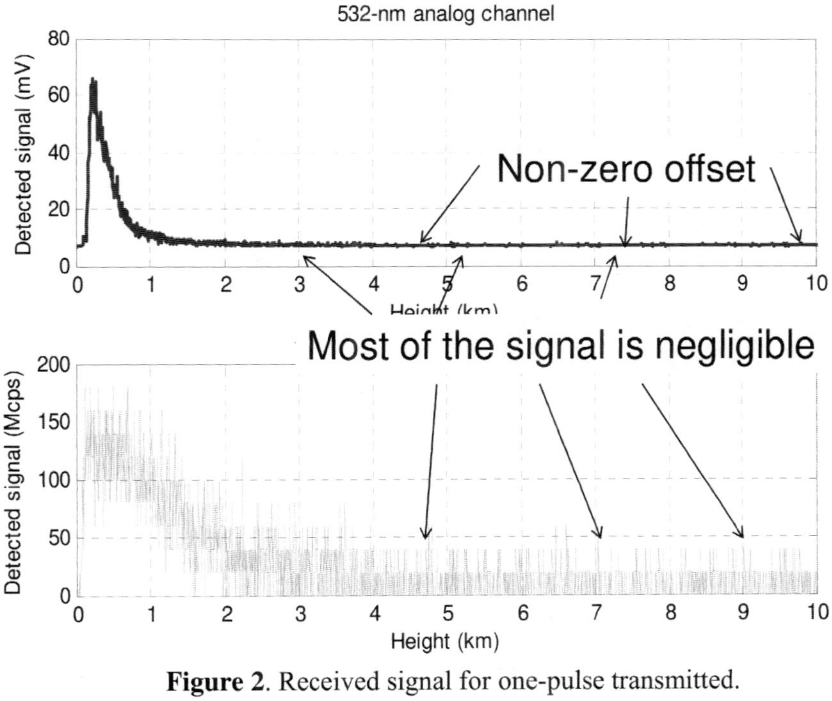

Figure 2. Received signal for one-pulse transmitted.

The received signal power ($P(R)$) contains information about the atmospheric back-scattering ($\beta(R)$) and extinction ($\alpha(R)$). It also states that the received signal is inversely proportional to the squared distance. The rest of the parameters are: E_{pulse}, the transmitter laser pulse energy and A_{teles}, the collecting area of the receiving telescope. In order to obtain a useful signal we will apply offset subtracting, 150-minute averaging and range correction (multiplying the signal times the squared distance). This measurement is one of the regular measurements which our group performs simultaneously with close passes of CALIPSO satellite [3].The result of the previous processes can be seen in Figure 3.

3. Dead-time correction

From the previous section, it must be pointed out that both (analog and photon-counting) channels present low signal levels at short distance. In the photon-counting channel, dead-time effect (an event will not be recorded when the time interval between its occurrence and the occurrence of the last recorded event is smaller than some given interval, τ_{dead}) must be corrected. A good estimate of the true number of the collected photons can be obtained with the equation [2]:

$$N_{real} = \frac{N_{meas}}{1-\tau_{dead} N_{meas}} \quad (2)$$

The result of applying this relation is plotted in Figure 4.

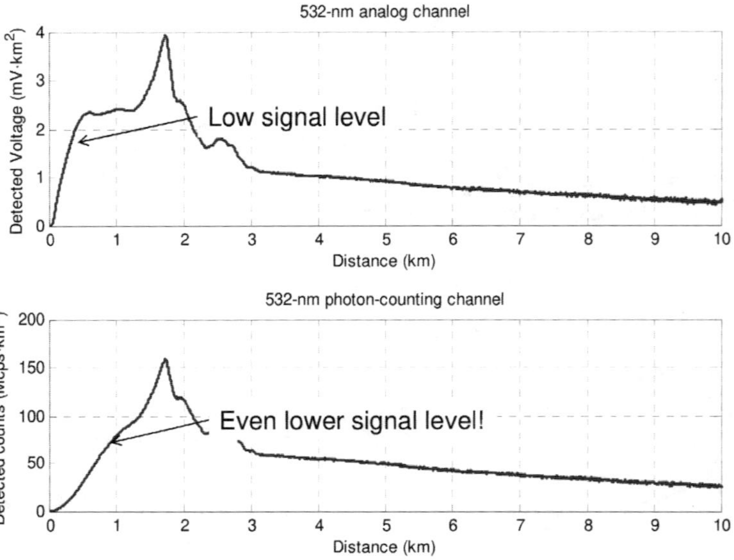

Figure 3. Time-averaged and ranged corrected signals.

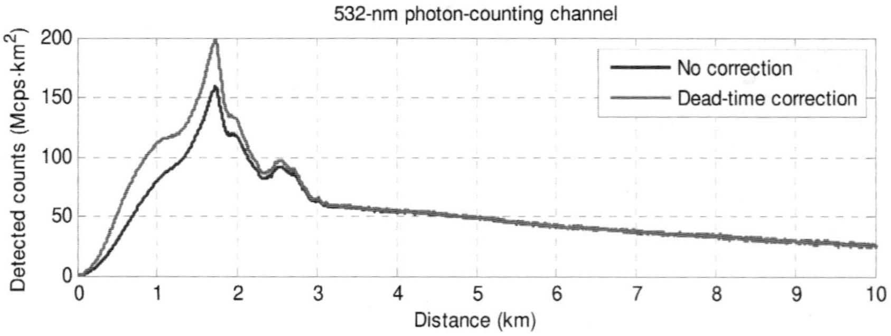

Figure 4. Dead-time correction.

4. Overlap correction

Even though it is not indicated in Figure 1, there usually is a non-negligible distance between the transmitter and the receiving telescope axes. There could also be some kind of misalignment between both axes. The effect of this separation is wonderfully explained in [1]. Figure 5 shows the effect of the separation of both axes, together with the actual values of laser beam divergence and receiver field of view.

For these values, the effect of the incomplete overlap can be either computed theoretically or estimated from measurements. Wandinger and Ansmann [5] demonstrated that the effect of the overlap in a particular measurement can be obtained by comparing the KFS inversion (affected by the overlap effect) and the Raman inversion (not affected). The result of the comparison (which involves techniques out the scope of this paper) is shown in Figure 6. Overlap is computed from 100 to 1000 m. Then it can be applied to correct both analog and photon-counting channels. The effect of this correction is plotted in Figure 7. It must be noted that, as expected, the result with the analog channel is a stronger return from low altitudes, where the concentration of aerosols is usually highest. Nevertheless, the low panel shows still low values for the photon-counting channel: these low values are usually attributed to saturation of the event counter when many photons are available. This saturation cannot be corrected reliably. Thus, some combination between the analog and the photon-counting signals must be performed, as it will be described in Section 6.

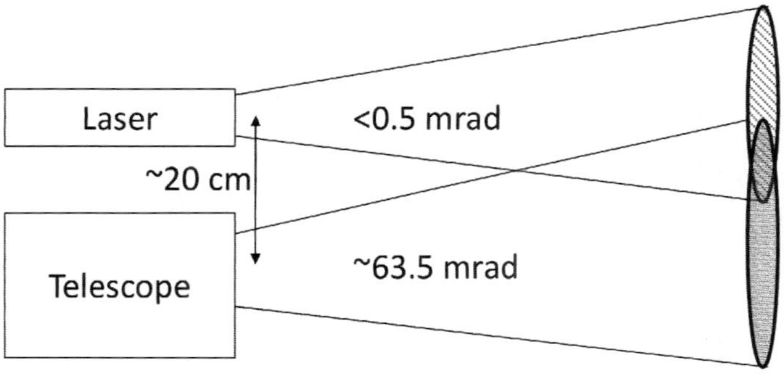

Figure 5. Illustration of the effect of partial overlap between the transmitted beam and the receiver field of view (adapted from [1]) .

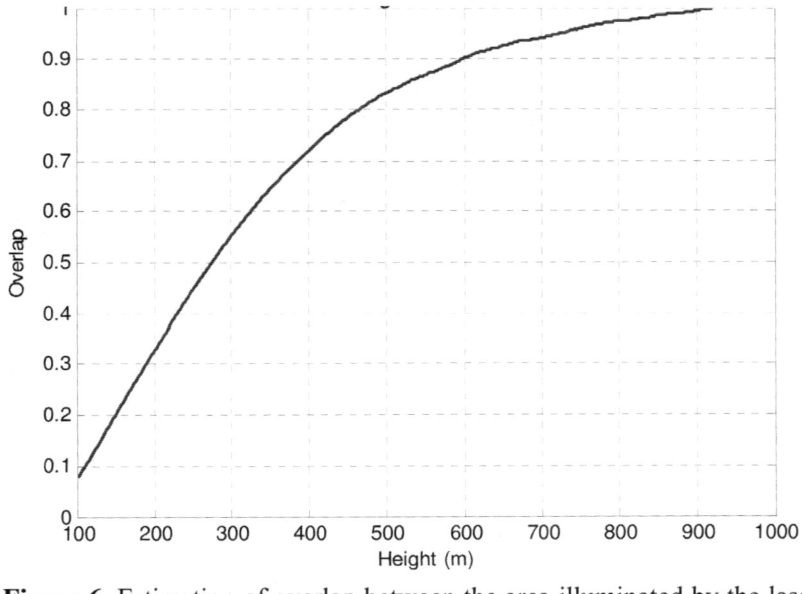

Figure 6. Estimation of overlap between the area illuminated by the laser and that seen by the lidar receiver.

5. Rayleigh fit

In all the plots shown before, it can be seen that neither the analog nor the photon-counting signals reach a zero-value, which is not consistent with the fact that aerosols are found only in the first kilometres of the troposphere. To get an answer to this fact, one has to remember that two light-

IOP Conf. Series: Earth and Environmental Science **28** (2015) 012013 doi:10.1088/1755-1315/28/1/012013

scattering mechanisms are present in the atmosphere: Mie scattering, produced by aerosols (roughly of the same size as the light wavelength) and Rayleigh scattering, produced by atmospheric constituent gases. With this in mind, we should rewrite equation (1) in the form:

$$P(R)=\frac{E_{pulse}A_{telescope}c}{2R^2}\left[\beta_{aer}(R)+\beta_{gases}(R)\right]\cdot e^{-2\int_0^R\left[\alpha_{aer}(x)+\alpha_{gases}(x)\right]dx} \quad (3)$$

So, the next step in the lidar signal pre-processing will be to identify the component produced by atmospheric gases in our signal and subtract it from the detected signal. The gas component can be either computed from standard atmospheric models [6] or from measurements obtained by "radio-sounds".

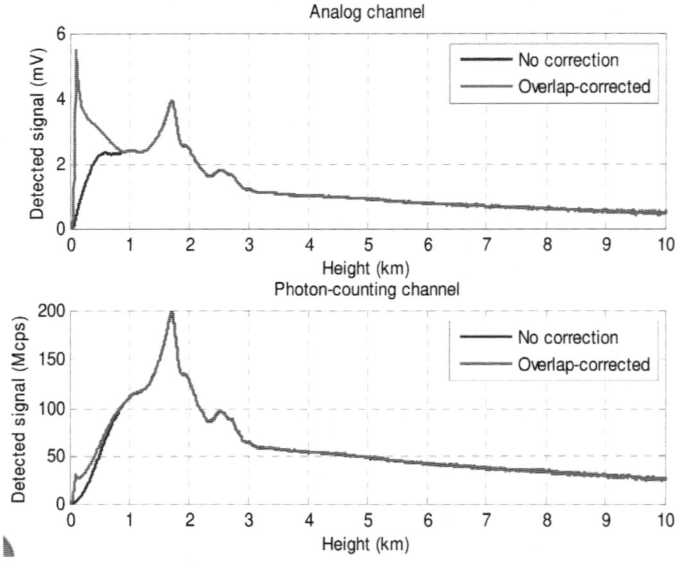

Figure 7. Effect of overlap correction.

Radio-sound measurements are performed in Barcelona by the Catalan Meteorological Service twice a day, at 00:20 and 12:20 UTC. So the lidar systems can be fitted by using a least squared error approach in the range of height values where no aerosol is expected. Figure 8 shows both the Rayleigh profiles computed from radio-sound measurements and the lidar signals fitted to the Rayleigh profile in the height interval between 6 and 10 km.

6. Gluing

Along the different plots presented in this paper, the analog channel represents more accurately the aerosol return at low altitude. Although it might not be evident, the photon-counting channel provides better Signal-to-Noise ratio at high altitude. For this reason, the two signals are combined in a process called gluing. Several strategies have been proposed [4] but we follow that proposed by Diego Lange in his PhD dissertation [7]. The result is presented in Figure 9.

7. Aerosol-only return

Finally, the signal due only to aerosol return can be identified and fed to the inversion algorithms, by subtracting the gas return (green line in Figure 9) from the composed signal (blue line in Figure 9). The result is plotted in Figure 10.

IOP Conf. Series: Earth and Environmental Science **28** (2015) 012013 doi:10.1088/1755-1315/28/1/012013

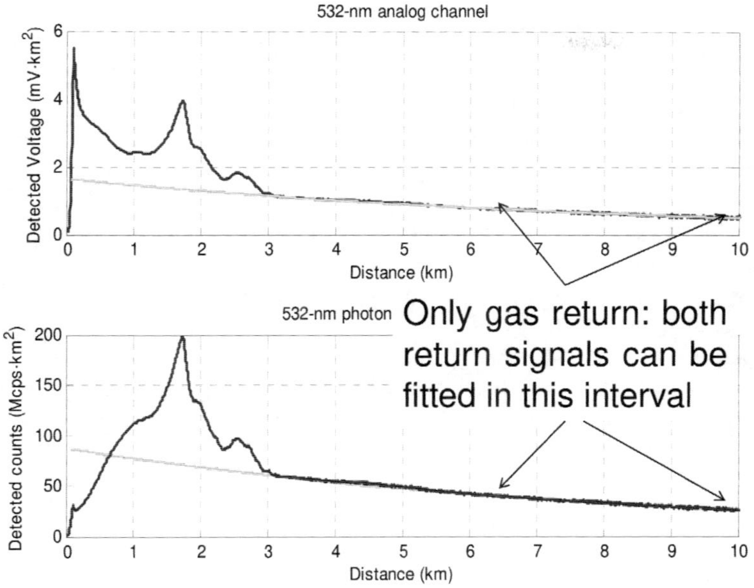

Figure 8. Rayleigh profiles and Rayleigh-fitted signals from 6 to 10 km.

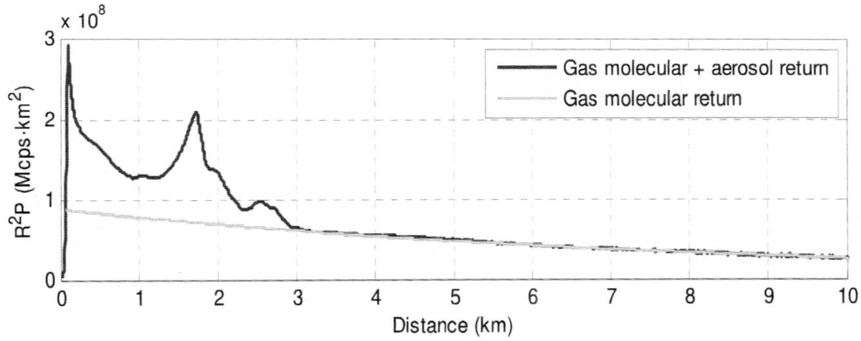

Figure 9. Resulting signal from gluing analog and photon-counting channels.

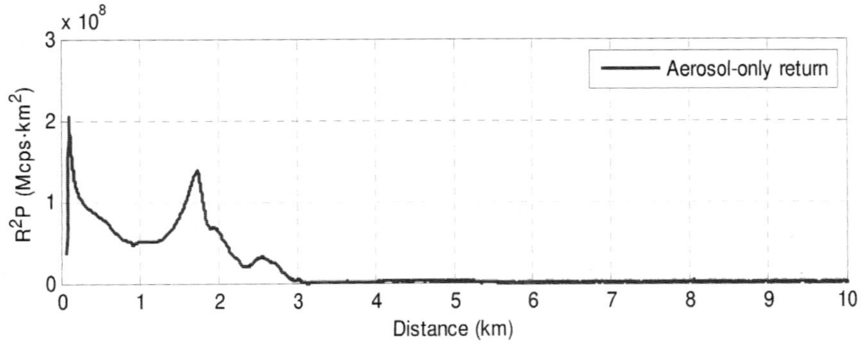

Figure 10. Aerosol-only return signal.

Acknowledgements
This work has been financed by Spanish Ministry for Economy and Competitiveness, through projects TEC2012-34799 and TEC2012-34575.

IOP Conf. Series: Earth and Environmental Science **28** (2015) 012013 doi:10.1088/1755-1315/28/1/012013

References

[1] Measures R, Laser Remote Sensing: Fundamentals and Applications, Krieger Publishing Company 1992

[2] Kumar D, Rocadenbosch F, Concept Design, Analysis, and Integration of the New UPC Multispectral Lidar System, UPC 2012

[3] http://www-calipso.larc.nasa.gov/about/payload.php

[4] Licel GmbH, Licel Ethernet Controller – Installation and Reference Manual.

[5] Wandinger U, Ansmann A, "Experimental determination of the lidar overlap profile with Raman lidar," *Appl. Opt.* 41, 511-514 (2002).

[6] http://www.atmosculator.com/The%20Standard%20Atmosphere.html

[7] Lange D, Rocadenbosch F, Lidar and S-band radar profiling of the atmosphere: Adaptive processing for Boundary-Layer monitoring, optical-parameter error estimation, and application cases, Universitat Politècnica de Catalunya-BarcelonaTech, 2014.

Institute of Physics
Dirac House, Temple Back
Bristol BS1 6BE UK

ISSN: 1755-1307
ISBN 978-1-5108-1633-6

9 781510 816336

CURRAN ASSOCIATES INC.
proceedings.com